Bioinformatics: Concepts and Applications

Bioinformatics: Concepts and Applications

Lawrence Baker

Larsen & Keller
www.larsen-keller.com

Bioinformatics: Concepts and Applications
Lawrence Baker
ISBN: 978-1-64172-105-9 (Hardback)

☰ Larsen & Keller

Published by Larsen and Keller Education,
5 Penn Plaza,
19th Floor,
New York, NY 10001, USA

Cataloging-in-Publication Data

Bioinformatics : concepts and applications / Lawrence Baker.
    p. cm.
Includes bibliographical references and index.
ISBN 978-1-64172-105-9
1. Bioinformatics. 2. Biology--Data processing. I. Baker, Lawrence.
QH324.2 .B56 2019
570--dc23

For more information regarding Larsen and Keller Education and its products, please visit the publisher's website www.larsen-keller.com

Table of Contents

Permissions

Index

Preface

Bioinformatics is an interdisciplinary science that develops on the methods and principles of statistics, computing, mathematics and biology to analyze biological data. It also includes the study of protein structures, amino acid sequences and nucleotide sequences. Techniques such as machine learning algorithms, pattern recognition, data mining and visualization are used. Drug discovery and design, gene finding, sequence alignment, protein-protein interactions, etc. are important areas of interest. This book aims to shed light on some of the unexplored aspects of bioinformatics. It elucidates new techniques and their applications in a multidisciplinary approach. In this book, constant effort has been made to make the understanding of the difficult concepts of bioinformatics as easy and informative as possible, for the readers.

To facilitate a deeper understanding of the contents of this book a short introduction of every chapter is written below:

Chapter 1, Bioinformatics is a field that seeks to understand biological data through the development of software tools. This chapter provides an overview of the basics of bioinformatics and biological data. It includes vital topics related to varied aspects of biological databases, model organism databases, sequence databases, gene disease databases, etc. **Chapter 2**, In bioinformatics, the sequence of DNA, RNA and protein are arranged for identifying regions of similarity, based on structural, functional or evolutionary relationships. This process is termed as sequence alignment. The aim of this chapter is to explore the crucial aspects of biological sequence alignment, alignment-free sequence analysis and biological sequence format. **Chapter 3**, Determining the precise order of nucleotides in a DNA molecule is called DNA sequencing. The investigation of the presence and quantity of RNA at a given moment in a biological sample is done using RNA sequencing. This chapter has been carefully written to provide an easy understanding of the varied aspects of DNA patterns, RNA and DNA sequencing. **Chapter 4**, Bioinformatics involves the development of algorithms, databases, and statistical and computational techniques for the solution of problems in the analysis of biological data. The aim of this chapter is to explore the varied algorithms used in bioinformatics, such as Baum–Welch algorithm, Needleman–Wunsch algorithm, String searching algorithm, Hirschberg's algorithm, Smith–Waterman algorithm, etc. which are crucial for a complete understanding of bioinformatics. **Chapter 5**, The software used in bioinformatics range in complexity from simple command-line tools to complex graphical programs. This chapter explores the different software used in bioinformatics such as FASTA, Clustal, GeneMark, GenoCAD, RAPTOR, sequence profiling tool, bowtie, etc. **Chapter 6**, Bioinformatics is a vast subject that branches out into a number of significant sub-disciplines such as integrative bioinformatics, neuroinformatics, glycoinformatics, bioimage informatics, neighbor joining, etc. These sub-disciplines have been thoroughly discussed in this chapter.

I would like to share the credit of this book with my editorial team who worked tirelessly on this book. I owe the completion of this book to the never-ending support of my family, who supported me throughout the project.

Lawrence Baker

Bioinformatics: Understanding Biological Data

Bioinformatics is a field that seeks to understand biological data through the development of software tools. This chapter provides an overview of the basics of bioinformatics and biological data. It includes vital topics related to varied aspects of biological databases, model organism databases, sequence databases, gene disease databases, etc.

Bioinformatics involves the integration of computers, software tools, and databases in an effort to address biological questions. Bioinformatics approaches are often used for major initiatives that generate large data sets. Two important large-scale activities that use bioinformatics are genomics and proteomics. Genomics refers to the analysis of genomes. A genome can be thought of as the complete set of DNA sequences that codes for the hereditary material that is passed on from generation to generation. These DNA sequences include all of the genes (the functional and physical unit of heredity passed from parent to offspring) and transcripts (the RNA copies that are the initial step in decoding the genetic information) included within the genome. Thus, genomics refers to the sequencing and analysis of all of these genomic entities, including genes and transcripts, in an organism. Proteomics, on the other hand, refers to the analysis of the complete set of proteins or proteome. In addition to genomics and proteomics, there are many more areas of biology where bioinformatics is being applied (i.e., metabolomics, transcriptomics). Each of these important areas in bioinformatics aims to understand complex biological systems.

Many scientists today refer to the next wave in bioinformatics as systems biology, an approach to tackle new and complex biological questions. Systems biology involves the integration of genomics, proteomics, and bioinformatics information to create a whole system view of a biological entity.

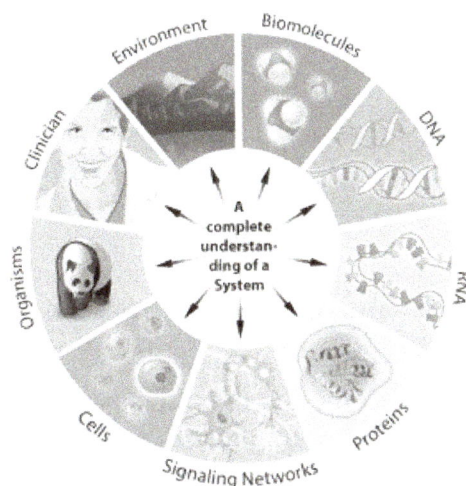

The Wheel of Biological Understanding. System biology strives to understand all aspects of an organism and its environment through the combination of a variety of scientific fields.

For instance, how a signaling pathway works in a cell can be addressed through systems biology. The genes involved in the pathway, how they interact, and how modifications change the outcomes downstream, can all be modeled using systems biology. Any system where the information can be represented digitally offers a potential application for bioinformatics. Thus bioinformatics can be applied from single cells to whole ecosystems. By understanding the complete "parts lists" in a genome, scientists are gaining a better understanding of complex biological systems. Understanding the interactions that occur between all of these parts in a genome or proteome represents the next level of complexity in the system. Through these approaches, bioinformatics has the potential to offer key insights into our understanding and modeling of how specific human diseases or healthy states manifest themselves.

The beginning of bioinformatics can be traced back to Margaret Dayhoff in 1968 and her collection of protein sequences known as the Atlas of Protein Sequence and Structure. One of the early significant experiments in bioinformatics was the application of a sequence similarity searching program to the identification of the origins of a viral gene. In this study, scientists used one of the first sequence similarity searching computer programs (called FASTP), to determine that the contents of v-sis, a cancer-causing viral sequence, were most similar to the well-characterized cellular PDGF gene. This surprising result provided important mechanistic insights for biologists working on how this viral sequence causes cancer. From this first initial application of computers to biology, the field of bioinformatics has exploded. The growth of bioinformatics is parallel to the development of DNA sequencing technology. In the same way that the development of the microscope in the late 1600's revolutionized biological sciences by allowing Anton Van Leeuwenhoek to look at cells for the first time, DNA sequencing technology has revolutionized the field of bioinformatics. The rapid growth of bioinformatics can be illustrated by the growth of DNA sequences contained in the public repository of nucleotide sequences called GenBank.

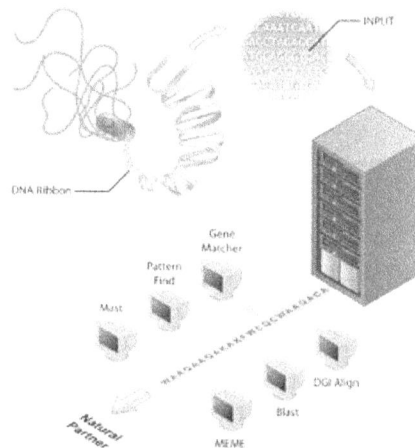

The Use of Computers to Process Biological Information. The wealth of genome sequencing information has required the design of software and the use of computers to process this information.

Genome sequencing projects have become the flagships of many bioinformatics initiatives. The human genome sequencing project is an example of a successful genome sequencing project but many other genomes have also been sequenced and are being sequenced. In fact, the first genomes to be sequenced were of viruses (i.e., the phage MS2) and bacteria, with the genome of Haemophilus influenzae Rd being the first genome of a free living organism to be deposited into the public sequence databanks. This accomplishment was received with less fanfare than the completion of the human genome but it is becoming clear that the sequencing of other genomes is an important step for bioinformatics today. However, genome sequence by itself has limited information. To interpret genomic information, comparative analysis of sequences needs to be done and an important reagent for these analyses are the publicly accessible sequence databases. Without the databases of sequences (such as GenBank), in which biologists have captured information about their sequence of interest, much of the rich information obtained from genome sequencing projects would not be available.

The same way developments in microscopy foreshadowed discoveries in cell biology, new discoveries in information technology and molecular biology are foreshadowing discoveries in bioinformatics. In fact, an important part of the field of bioinformatics is the development of new technology that enables the science of bioinformatics to proceed at a very fast pace. On the computer side, the Internet, new software developments, new algorithms, and the development of computer cluster technology has enabled bioinformatics to make great leaps in terms of the amount of data which can be efficiently analyzed. On the laboratory side, new technologies and methods such as DNA sequencing, serial analysis of gene expression (SAGE), microarrays, and new mass spectrometry chemistries have developed at an equally blistering pace enabling scientists to produce data for analyses at an incredible rate. Bioinformatics provides both the platform technologies that enable scientists to deal with the large amounts of data produced through genomics and proteomics initiatives as well as the approach to interpret these data. In many ways, bioinformatics provides the tools for applying scientific method to large-scale data and should be seen as a scientific approach for asking many new and different types of biological questions.

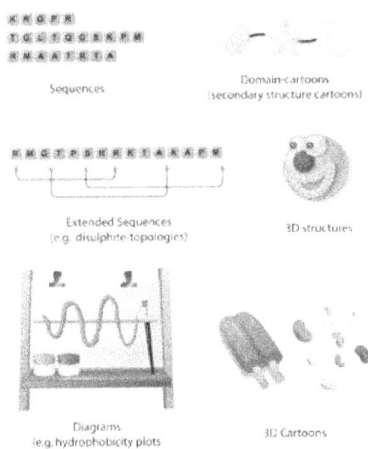

Potential Types of Bioinformatic Data. Computer based databases of biological information enables scientist to generate all sorts of data, from generating protein sequence and predicting protein domains to even producing 3D structures of proteins.

The word bioinformatics has become a very popular "buzz" word in science. Many scientists find bioinformatics exciting because it holds the potential to dive into a whole new world of uncharted territory. Bioinformatics is a new science and a new way of thinking that could potentially lead to many relevant biological discoveries. Although technology enables bioinformatics, bioinformatics is still very much about biology. Biological questions drive all bioinformatics experiments. Important biological questions can be addressed by bioinformatics and include understanding the genotype-phenotype connection for human disease, understanding structure to function relation-ships for proteins, and understanding biological networks. Bioinformaticians often find that the reagents necessary to answer these interesting biological questions do not exist. Thus, a large part of a bioinformatician's job is building tools and technologies as part of the process of asking the question. For many, bioinformatics is very popular because scientists can apply both their biology and computer skills to developing reagents for bioinformatics research. Many scientists are find-ing that bioinformatics is an exciting new territory of scientific questioning with great potential to benefit human health and society.

The future of bioinformatics is integration. For example, integration of a wide variety of data sourc-es such as clinical and genomic data will allow us to use disease symptoms to predict genetic muta-tions and vice versa. The integration of GIS data, such as maps, weather systems, with crop health and genotype data, will allow us to predict successful outcomes of agriculture experiments. Anoth-er future area of research in bioinformatics is large-scale comparative genomics. For example, the development of tools that can do 10-way comparisons of genomes will push forward the discovery rate in this field of bioinformatics. Along these lines, the modeling and visualization of full net-works of complex systems could be used in the future to predict how the system (or cell) reacts, to a drug, for example. A technical set of challenges faces bioinformatics and is being addressed by faster computers, technological advances in disk storage space, and increased bandwidth, but by far one of the biggest hurdles facing bioinformatics today, is the small number of researchers in the field. This is changing as bioinformatics moves to the forefront of research but this lag in exper-tise has lead to real gaps in the knowledge of bioinformatics in the research community. Finally, a key research question for the future of bioinformatics will be how to computationally compare complex biological observations, such as gene expression patterns and protein networks. Bioin-formatics is about converting biological observations to a model that a computer will understand. This is a very challenging task since biology can be very complex. This problem of how to digitize phenotypic data such as behavior, electrocardiograms, and crop health into a computer readable form offers exciting challenges for future bioinformaticians.

Goals of Bioinformatics

The development of efficient algorithms for measuring sequence similarity is an important goal of bio-informatics. The Needleman-Wunsch algorithm, which is based on dynamic programming, guarantees finding the optimal alignment of pairs of sequences. This algorithm essentially divides a large problem (the full sequence) into a series of smaller problems (short sequence segments) and uses the solu-tions of the smaller problems to construct a solution to the large problem. Similarities in sequences are scored in a matrix, and the algorithm allows for the detection of gaps in sequence alignment.

Although the Needleman-Wunsch algorithm is effective, it is too slow for probing a large sequence database. Therefore, much attention has been given to finding fast information-retrieval algo-rithms that can deal with the vast amounts of data in the archives. An example is the program BLAST (Basic Local Alignment Search Tool). A development of BLAST, known as position-specific iterated- (or PSI-) BLAST, makes use of patterns of conservation in related sequences and com-bines the high speed of BLAST with very high sensitivity to find related sequences.

Another goal of bioinformatics is the extension of experimental data by predictions. A fundamental goal of computational biology is the prediction of protein structure from an amino acid sequence.

The spontaneous folding of proteins shows that this should be possible. Progress in the development of methods to predict protein folding is measured by biennial Critical Assessment of Structure Prediction (CASP) programs, which involve blind tests of structure prediction methods.

Bioinformatics is also used to predict interactions between proteins, given individual structures of the partners. This is known as the "docking problem." Protein-protein complexes show good complementarity in surface shape and polarity and are stabilized largely by weak interactions, such as burial of hydrophobic surface, hydrogen bonds, and van der Waals forces. Computer programs simulate these interactions to predict the optimal spatial relationship between binding partners. A particular challenge, one that could have important therapeutic applications, is to design an antibody that binds with high affinity to a target protein.

Initially, much bioinformatics research has had a relatively narrow focus, concentrating on devising algorithms for analyzing particular types of data, such as gene sequences or protein structures. Now, however, the goals of bioinformatics are integrative and are aimed at figuring out how combinations of different types of data can be used to understand natural phenomena, including organisms and disease.

Biological Databases

Biological databases emerged as a response to the huge data generated by low-cost DNA sequencing technologies. One of the first databases to emerge was GenBank, which is a collection of all available protein and DNA sequences. It is maintained by the National Institutes of Health (NIH) and the National Center for Biotechnology Information (NCBI). GenBank paved the way for the Human Genome Project (HGP). The HGP allowed complete sequencing and reading of the genetic blueprint. The data stored in biological databases is organized for optimal analysis and consists of two types: raw and curated (or annotated). Biological databases are complex, heterogeneous, dynamic, and yet inconsistent. The inconsistency is due to the lack of standards at the ontological level.

Importance of Biological Database

Earlier, databases and databanks were considered quite different. However, over the time, database became a preferable term. Data is submitted directly to biological databases for indexing, organization, and data optimization. They help researchers find relevant biological data by making it available in a format that is readable on a computer. All biological information is readily accessible through data mining tools that save time and resources. Biological databases can be broadly classified as sequence and structure databases. Structure databases are for protein structures, while sequence databases are for nucleic acid and protein sequences.

Kinds of Biological Databases

Biological databases can be further classified as primary, secondary, and composite databases.

Primary databases contain information for sequence or structure only. Examples of primary biological databases include:

- Swiss-Prot and PIR for protein sequences

- GenBank and DDBJ for genome sequences

- Protein Databank for protein structures

Secondary databases contain information derived from primary databases. Secondary databases store information such as conserved sequences, active site residues, and signature sequences. Protein Databank data is stored in secondary databases. Examples include:

- SCOP at Cambridge University

- CATH at the University College of London

- PROSITE of the Swiss Institute of Bioinformatics

- eMOTIF at Stanford

Composite databases contain a variety of primary databases, which eliminates the need to search each one separately. Each composite database has different search algorithms and data structures. The NCBI hosts these databases, where links to the Online Mendelian Inheritance in Man (OMIM) is found.

Sequence Databases

A sequence database is a collection of DNA or protein sequences with some extra relevant information. The main sequence databases are Genbank and EMBL. Originally they were just sequence collections, but they have grown to store different biological databases heavily interconnected and they provide powerful interfaces to search and browse the stored information.

The sequences submitted to any of those databases are shared between them, so any sequence could be retrieved in the european or the american database. But they differ in the tools to search and browse the data and in some databases that provide extra information to the raw sequences like: mutations, coded proteins, bibliographical references, etc.

These databases are growing at an ever increasing fast pace. In June of 2007 there were 73 million sequences in Genbank and in August of 2015 there were 187 millions.

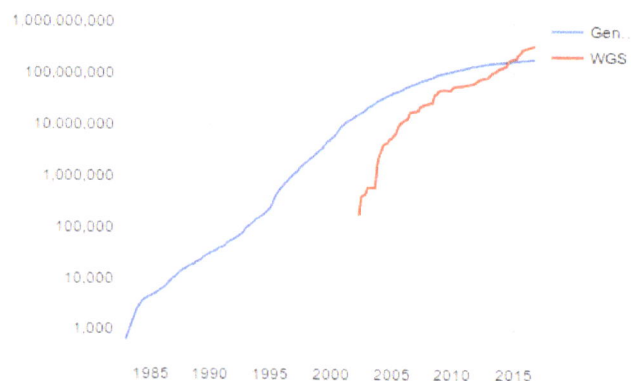

The sequences are split in these databases in different sections to ease the search. Among others, there are sections for mRNAs, publised nucleotide sequences, genomes, and genes.

Genbank

Genbank is a public collection of annotated sequences hosted by the NCBI. Among other kinds of sequences Genbank includes messenger RNAs, genomic DNAs and ribosomic RNA.

Some characteristics:

- It is a public repository, any one can send sequences to it.

- There are sequences of different qualities, anything submitted is stored.

- There could be multiple sequences for the same gene or for the same mRNA.

- A sequence can have several versions that represent the modifications done by the authors.

Due to the huge amount of sequences stored to ease the search the databases are split in different divisions. These divisions follow two criteria: the species and type of sequence. Among the taxonomical divisions you can find: primate, rodent, other mammalian, invertebrate an others. The other divisions are related to the kind of sequences like: EST, WGS, HTGS, and many others. If you are looking for reads comming from the Next Generation Sequencing Technologies they are stored in a special division called SRA.

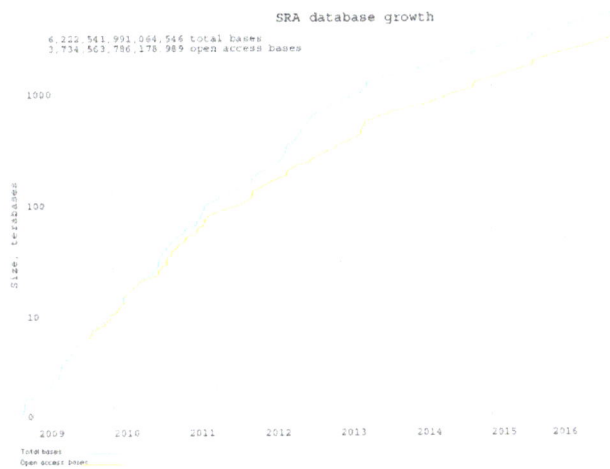

RefSeq

RefSeq is a reference database curated by NCBI.

In RefSeq there are only well annotated and good quality sequences. It stores genomic, transcript and protein sequences and links the sequences that belong to a gene. It just has one representative sequence for each mRNA in a particular organism and, thus, it will have as many sequences as different transcripts and proteins coded for a particular gene in a particular organism.

It is not the aim of ReqSeq to have any sequence, but just to have a collection of well curated sequences. It is a secondary database. Since RefSeq requires extra curation work it is not available for all organisms, but only for those with good quality sequences. As of July of 2016 it had 65M proteins and 15M transcripts for 60K organisms.

UniProt

UniProt is a protein database that includes information divided in two sections: Swiss-Prot and TrEMBL. UniProt aims to store sequence and functional information for the proteins.

TrEMBL is automatically annotated while Swiss-Prot is reviewed manually by humans that add information by reviewing the literature. Due to this effort Swiss-Prot has information of a higher quality, but it has less sequences than TrEMBL.

UniProt also hosts Uniref. This database aims to store one representative sequence for each protein without taking into account the species of origin. It clusters all the similar proteins and picks one for every cluster as a representative. There are clusters created at 100%, 90% and 50% identities.

PubMed

PubMed is a bibliographical database that comprises biomedical literature (MEDLINE), life science journals and on-line books. It is a good collection of publications related to biochemistry, cellular biology and medicine. As of 2016 PubMed stored 26 million citations.

For each record it stores:

- title

- authors

- abstract

There is a related database named PubMed Cental (PMC) that only includes citations of Free Access Journals. These citations include the complete text for the papers stored.

PDB, Protein Data Bank

PDB stores 3D structures for proteins and nucleic acids.

Access to the Information in Genbank

Every database provides one or more methods to search and query the data. It is quite common to provide a web interface in which to do text searches with some keyword, author, ID or any other text. Genbank has a powerful query web interface.

Search NCBI databases

[Search]

Literature		Genes	
Books	books and reports	EST	expressed sequence tag sequences
MeSH	ontology used for PubMed indexing	Gene	collected information about gene loci
NLM Catalog	books, journals and more in the NLM Collections	GEO DataSets	functional genomics studies

Each database shows the results in one or several formats. For instance, the Genbank sequences can be obtained in several formats.

Genbank format:

```
LOCUS       EC750390            558 bp  mRNA   linear   EST   03-JUL-2006

DEFINITION POE00005652 PL(light) Polytomella parva cDNA similar to frataxin pro-
tein

        -related, mRNA sequence.

ACCESSION   EC750390

VERSION     EC750390.1 GI:110064507

KEYWORDS    EST.

SOURCE    Polytomella parva

ORGANISM  Polytomella parva

      Eukaryota; Viridiplantae; Chlorophyta; Chlorophyceae;Chlamydomonadales;

        Chlamydomonadaceae; Polytomella.

REFERENCE   1 (bases 1 to 558)

 AUTHORS   Lee,R.W. and Borza,T.

 TITLE    The colorless plastid of the green alga Polytomella parva: a repertoire
of its functions

 JOURNAL   Unpublished (2006)

COMMENT    Contact: TBestDB

      Departement de Biochimie, Universite de Montreal

      Montreal, Canada

      Email: tbestdb-curator@bch.umontreal.ca

      Plate: 4065.

FEATURES          Location/Qualifiers

   source    1..558

          /organism="Polytomella parva"

          /mol_type="mRNA"

          /db_xref="taxon:51329"

          /clone_lib="PL(light)"

ORIGIN

    1 gcggccgctt tttttttttt tttttttttt ttttcgtccg ttatttcttt tttaagaatg

   61 cagtcatctg tacatcgtca agtattcgga gtgttatctc gttttgtggg aaacaaagcg
```

```
121 ggtatttta caaagcataa tcatggtgtc tcaaggttgt cttcatgcac ttcgtcatgc

181 gtaaagatgt atactagcaa caaggccccc gaggatcttc aaacgttcca ccggcaagca

241 gacgaaactc tagagcaagt cactgaagcc cttgaaaact atgtagatga gcatgaagtg

301 gaaggcagcg acattgagca tacgcaagga gtgcttacta ttaagcttgg aactcttgga

361 agttatgtaa ttaataaaca gactcctaat aagcagatat ggttatcctc tcccgtcagt

421 ggacccttcc gatatgatct aaagaaggt gcctgggttt atgaacgggc tggcgaggct

481 cggcgcgagc ttatttctca attagaaaca gaaatttcgg atttagttgg tgtcgaatta

541 aagataagta actgaacg
```

EMBL format:

```
ID  EC750390; SV 1; linear; mRNA; EST; PLN; 558 BP.

XX

AC  EC750390;

XX

DT  04-JUL-2006 (Rel. 88, Created)

DT  04-JUL-2006 (Rel. 88, Last updated, Version 1)

XX

DE  POE00005652 PL(light) Polytomella parva cDNA similar to frataxin

DE  protein-related, mRNA sequence.

XX

KW  EST.

XX

OS  Polytomella parva

OC  Eukaryota; Viridiplantae; Chlorophyta; Chlorophyceae; Chlamydomonadales;

OC  Chlamydomonadaceae; Polytomella.

XX

RN  [1]

RP  1-558

RA  Lee R.W., Borza T.;

RT  "The colorless plastid of the green alga Polytomella parva: a repertoire of

RT  its functions";
```

RL   Unpublished.

XX

DR   UNILIB; 42732; 19932.

XX

CC   Contact: TBestDB

CC   Departement de Biochimie, Universite de Montreal

CC   Montreal, Canada

CC   Email: tbestdb-curator@bch.umontreal.ca

CC   Plate: 4065.

XX

FH   Key   Location/Qualifiers

FH

FT   source   1..558

FT   /organism="Polytomella parva"

FT   /mol_type="mRNA"

FT   /clone_lib="PL(light)"

FT   /db_xref="taxon:51329"

FT   /db_xref="UNILIB:42732"

XX

SQ   Sequence 558 BP; 153 A; 105 C; 127 G; 173 T; 0 other;

     gcggccgctt tttttttttt tttttttttt ttttcgtccg ttatttcttt tttaagaatg
60

     cagtcatctg tacatcgtca agtattcgga gtgttatctc gttttgtggg aaacaaagcg
120

     ggtattttta caaagcataa tcatggtgtc tcaaggttgt cttcatgcac ttcgtcatgc
180

     gtaaagatgt atactagcaa caaggccccc gaggatcttc aaacgttcca ccggcaagca
240

     gacgaaactc tagagcaagt cactgaagcc cttgaaaact atgtagatga gcatgaagtg
300

```
    gaaggcagcg acattgagca tacgcaagga gtgcttacta ttaagcttgg aactcttgga
360
    agttatgtaa ttaataaaca gactcctaat aagcagatat ggttatcctc tcccgtcagt
420
    ggacccttcc gatatgatct taaagaaggt gcctgggttt atgaacgggc tggcgaggct
480
    cggcgcgagc ttatttctca attagaaaca gaaatttcgg atttagttgg tgtcgaatta
540
    aagataagta actgaacg
558
```

Main fields in the Genbank format

Field	Description	Search in Entrez
Locus name	Unique sequence name	[ACCN]
Sequence length	Sequence Length	[SLEN]
Molecule Type	DNA, genomic, mRNA, etc.	[PROP]
Genbank Division	Division for the sequence	[PROP]
Modification Date	Date for the last edit	[MDAT]
Definition	Brief description	[TITL]
Accession	Unique accession ID. It does not changes with modifications	[ACCN]
Version	Version number of the sequence	All fields
Keywords	keywords that describe the sequence	[KYWD]
Source	Common name for the source species	[ORGN]
Organism	Oficial name for the source species	[ORGN]
Reference	Related publications	[TITL] [AUTH] [JOUR]
Features	Regions of interest	[FKEY]
CDS	Coding Sequence	[FKEY]

The Accession is the unique identifier for a sequence record. An accession number applies to the complete record and is usually a combination of a letter(s) and numbers, such as a single letter followed by five digits (e.g., U12345) or two letters followed by six digits (e.g., AF123456).

The records in GenBank can be updated by an author request, accession numbers do not change, even if information in the record is changed. So, a sequence can have several versions in GenBank. Version is an unique identifier that represents a single, specific sequence in the GenBank database. If there is any change to the sequence data (even a single base), the version number will be increased, e.g., U12345.1 → U12345.2, but the accession portion will remain stable.

Features holds information about genes and gene products, as well as regions of biological significance reported in the sequence. These can include regions of the sequence that code for proteins and RNA molecules.

Sequence Formats

Text Files

There are different formats to store sequences in a text file. Text files should only include Plain text. Graphics or any other binary information are not allowed in text files.

Microsoft Word files are not text files, they are binary files that happen to represent documents. These documents can include text among many other things like images, charts or formats.

Sequences in Plain Files

We could store the sequence in a text file by just writing the sequence. This files would had to include only IUPAC characters.

```
ACAAGATGCCATTGTCCCCCGGCCTCCTGCTGCTGCTGCTCTCCGGGGCCACGGCCACCGCTGCCCTGCCCCTGGAG-
GGTAC

GGCCCCACCGGCCGAGACAGCGAGCATATGCAGGAAGCGGCAGGAATAAGGAAAAGCAGCCTCCTGACTTTCCTC-
GCTTGGT

AGTGGACCTCCCAGGCCAGTGCCGGGCCCCTCATAGGAGAGGAAGCTCGGGAGGTGGCCAGGCGGCAGGAAGGCG-
CACCCCC

ATCCGCGCGCCGGGACAGAATGCCCTGCAGGAACTTCTTCTGGAAGACCTTCTCCTCCTGCAAATAAAA
```

This kind of file is seldom used because it lacks any metadata to identify the sequence.

Fasta Format

The Fasta file includes a name for the sequence and, optionally, some description. The sequence should be preceded by a line that starts with the symbol >. The name will be written after that symbol. Spaces are not allowed in the sequence name. If there is a description it will be found after a space in the same line.

Several sequences can be included in the same file.

It is one of the most common formats.

```
>sequence1_name description

ACAAGATGCCATTGTCCCCCGGCCTCCTGCTGCTGCTGCTCTCCGGGGCCACGGCCACCGCTGCCCTGCC

CCTGGAGGGTGGCCCCACCGGCCGAGACAGCGAGCATATGCAGGAAGCGGCAGGAATAAGGAAAAGCAGC

CTCCTGACTTTCCTCGCTTGGTGGTTTGAGTGGACCTCCCAGGCCAGTGCCGGGCCCCTCATAGGAGAGG

AAGCTCGGGAGGTGGCCAGGCGGCAGGAAGGCGCACCCCCCCAGCAATCCGCGCGCCGGGACAGAATGCC
```

```
CTGCAGGAACTTCTTCTGGAAGACCTTCTCCTCCTGCAAATAAAACCTCACCCATGAATGCTCACGC

>sequence2_name description

ACAAGATGCCATTGTCCCCCGGCCTCCTGCTGCTGCTGCTCTCCGGGGCCACGGCCACCGCTGCCCTGCC

CCTGGAGGGTGGCCCCACCGGCCGAGACAGCGAGCATATGCAGGAAGCGGCAGGA
```

If we want to include more information we could use the GenBank or EMBL formats. It is also very common in the sequences that come directly from a sequencing machine to include the quality information, for that purpose the most common format is FASTQ.

Gene Disease Databases

Gene/Disease Specific Databases

Gene/disease specific databases are curated, online collections of information on genetic variations in a single gene, gene family or set of genes implicated in a single disease.

Also known as locus specific databases (LSDBs), they provide their information to the scientific community free of charge and with minimal restrictions on how their data can be used. Approximately 1800 gene/disease specific databases currently exist, gathering information on specific genes and diseases from the published literature, direct submission by researchers, and numerous other sources. Critically, these databases curate this information, ensuring that it is of a consistently high quality.

Need for Gene Disease Databases

Gene/disease specific databases have long been recognised as the best way to collect, organise and share information on genetic variation and their effect on patients. By limiting their scope to particular genes or diseases, the database curators, typically academic researchers whose works focus exclusively on the genes/diseases in question, are better able to assess the consequences of the sequence changes reported in the database. As next-generation sequencing becomes more prevalent, expert oversight of this variety will take on an increasing importance. Gene/disease specific databases are also more able to identify and collect the clinical data that are of most use to the researchers and clinicians working with these genes and disorders.

Access to complete, curated molecular and clinical information on the variations implicated in a particular disorder has a profound impact on the type of research that can be conducted into inherited diseases and will pave the way to more accurate diagnoses and treatments for these diseases.

Example of Gene Disease Database

DisGeNET

Biomedical sciences are facing an enormous increase of data available in public sources, not only in volume, but also in nature. Translational bioinformatics has emerged as a new field to transform the huge wealth of biomedical data into clinical actions using bioinformatic approaches. By the integrative exploitation of genomic, phenomic and environmental information, translational bioinformatics will enable a deeper understanding of disease mechanisms. In the pursuit to implement personalized medicine, the clinical practitioners will increasingly rely on informatic resources that aid in the exploration and interpretation of data on the genetic determinants of disease. The availability of both, comprehensive knowledge sources on disease genes and tools that allow their analysis and exploitation, should lay the basis to achieve this goal. Currently, there are several resources that cover different aspects of our current knowledge on the genetic basis of human diseases. DisGeNET is one of these resources, whose aims are to cover all disease areas (Mendelian, complex and environmental diseases), with special care on the integration and standardization of data, and to provide open access on knowledge of genes associated to human diseases.

'DisGeNET discovery platform' includes a new version of the database and, more importantly, a new set of 'analysis tools' to facilitate and foster the study of the molecular underpinning of human diseases. An important aspect of the DisGeNET toolkit is to support different types of users. Since the scientific literature represents a rich, up-to-date source of knowledge on disease genes, the database also includes gene-disease associations (GDAs) mined from MEDLINE via a NLP-based approach.

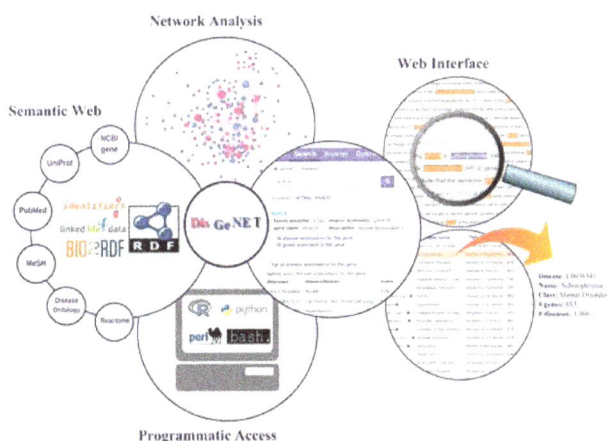

The main features of the DisGeNET discovery platform. DisGeNET is available through a web interface, a Cytoscape plugin, as a Semantic Web resource, and supports programmatic access to its data.

One of the key features of DisGeNET is the explicit representation of the provenance of the information, which allows the user to trace back to the original source of information and, more importantly, to explore the data in its original context. These aspects are of crucial importance to evaluate the evidence supporting a scientific assertion, in order to determine its relevance for translational applications. Moreover, the DisGeNET discovery platform allows prioritizing GDAs on the basis of the evidence supporting the data.

Semantic Web and Linked Data approaches have become increasingly important to life sciences and health care, since they properly meet the data standardization and integration requirements of translational biomedical research. The integration of DisGeNET in the emerging Semantic Web intends to ease and foster the integrated use of its data with other resources available in the web, and to support and expand research on human diseases and their genes.

The 'DisGeNET discovery platform' allows easy browsing and downloading of the information related to human diseases and their genes. The platform supports different types of users: the bioinformatician and software developer that interrogates the database by customized scripts or using Semantic Web technologies, the systems biology expert that explores and analyses the network representations of the information, and biologists and health-care practitioners who interrogate the database using its user-friendly web interface. Its comprehensiveness, standardization, availability and accessibility, as well as the suite of analysis tools and support of different user profiles make DisGeNET a resource of choice to investigate diseases of genetic origin.

The DisGeNET discovery platform

Challenges Creating Gene Disease Databases

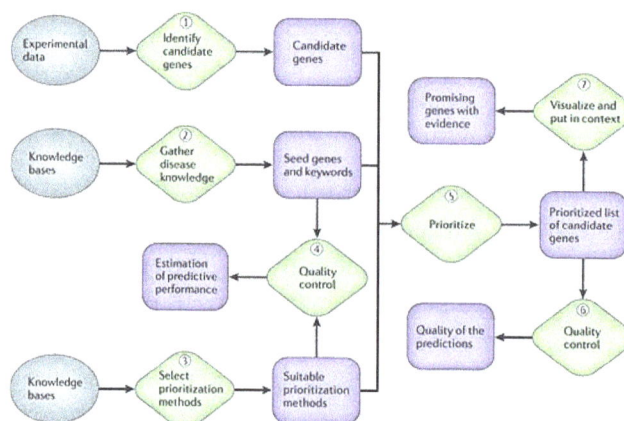

Gene prioritization workflow of human diseases

Typical lists come from linkage regions, chromosomal aberrations, association study loci, deferentially expressed gene lists or genes identified by sequencing variants. Alternatively, the complete genome can be prioritized, but substantially more false positives would then be expected.

At different stages of any gene disease project, molecular biologists need to choose, even after careful statistical data analysis, which genes or proteins to investigate further experimentally and which to leave out because of limited resources. Computational methods that integrate complex, heterogeneous data sets, such as expression data, sequence information, functional annotation and the biomedical literature, allow prioritizing genes for future study in a more informed way. Such methods can substantially increase the yield of downstream studies and are becoming invaluable to researchers. So one of the main concerns in biological and biomedical research is to recognise the underlying mechanisms behind this intricate genetic phenotypes. Great effort has been spent on finding the genes related to diseases

However, increasingly evidences point out that most human diseases cannot be attributed to a single gene but arise due to complex interactions among multiple genetic variants and environmental risk factors. Several databases have been developed storing associations between genes and diseases such as the Comparative Toxicogenomics Database (CTD), Online Mendelian Inheritance in Man (OMIM), the genetic Association Database (GAD) or the Disease genetic Association Database (DisGeNET). Each of these databases focuses on different aspects of the phenotype-genotype relationship, and due to the nature of the database curation process, they are not complete, but in a way they are fully complementary between each other.

Types of Databases

Essentially, there are four types of databases: curated databases, predictive databases, literature databases and integrative databases

Curated Databases

The term curated data refers to information, that may comprise the most sophisticated computational formats for structured data, scientific updates, and curated knowledge, that has been composed and prepared under the regulation of one or more experts considered to be qualified to engage in such an activity The implication is that the resulting database is of high quality. The contrast is with data which may have been gathered through some automated process or using particularly low or inexpert unsupported data quality and possibly untrustworthy. Some of the most common examples include: CTD and UNIPROT.

Comparative Toxicogenomics Database (CTD)

The Comparative Toxicogenomics Database, helps to understand about the effects of environmental compounds on human health by integrating data from curated scientific literature to describe biochemical interactions with genes and proteins, and links between diseases and chemicals, and diseases and genes or proteins. CTD contains curated data defining cross-species chemical–gene/protein interactions and chemical– and gene–disease associations to illuminate molecular mechanisms underlying variable susceptibility and environmentally influenced diseases. These data deliver insights into complex chemical–gene and protein interaction networks. One of the main sources in this Database is curated information from OMIM.

CTD is a unique resource where bioinformatics specialists read the scientific literature and manually curate four types of core data:

- Chemical-gene interactions

- Chemical-disease associations

- Gene-disease associations

- Chemical-phenotype associations

Universal Protein Resource (UNIPROT)

The Universal Protein Resource (UniProt) is an inclusive resource for protein sequence and annotation data. It is a comprehensive, first-class and freely accessible database of protein sequence and functional information, that has many entries being derived from genome sequencing projects. It contains a large amount of information about the biological function of proteins derived from the study literature, which can hint to a direct connection between gene-protein-disease.

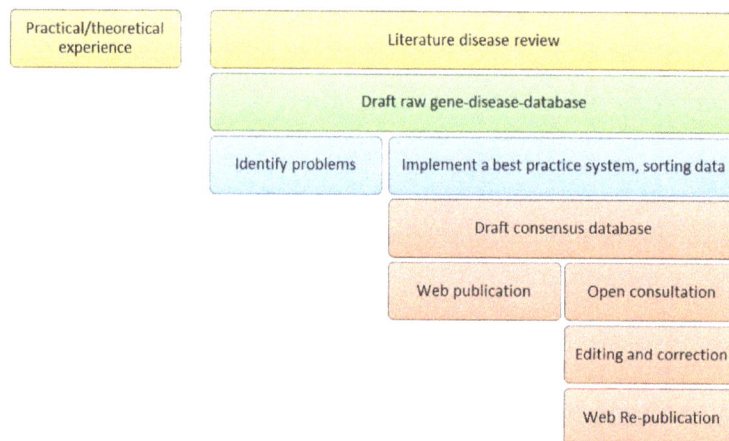

The process of database compilation and curation The curated data may comprise a process from practical experience and literature review to web publication of the database

Predictive Databases

A predictive database is one based on statistical inference. One particular approach to such inference is known as predictive inference, but the prediction can be undertaken within any of the several approaches to statistical inference. Indeed, one description of biostatistics is that it provides a means of transferring knowledge about a sample of a genetic population to the whole population (genomics), and to other related genes or genomes, which the same as prediction over time is not necessarily. When information is transferred across time, often to specific points in time, the process is known as forecasting. Three of the main examples of databases that can be considered in this category include: The Mouse genome Database (MGD), The Rat genome Database (RGD), OMIM and the SIFT Tool from Ensembl.

Mouse Genome Database

The Mouse genome Database (MGD) is the international community resource for integrated genetic, genomic and biological data about the laboratory mouse. MGD provides full annotation of phenotypes and human disease associations for mouse models (genotypes) using terms from the Mammalian Phenotype Ontology and disease names from OMIM.

Rat Genome Database (RGD)

The Rat Genome Database (RGD) began as a collaborative effort between leading research institutions involved in rat genetic and genomic research. The rat continues to be extensively used by researchers as a model organism for investigating the biology and pathophysiology of disease. In the past several years, there has been a rapid increase in rat genetic and genomic data. This explosion of information highlighted the need for a centralized database to efficiently and effectively collect, manage, and distribute a rat-centric view of this data to researchers around the world. The Rat Genome Database was created to serve as a repository of rat genetic and genomic data, as well as mapping, strain, and physiological information. It also facilitates investigators research efforts by providing tools to search, mine, and predict this data.

Data at RGD that is useful for researchers investigating disease genes include disease annotations for rat, mouse and human genes. Annotations are manually curated from the literature, or downloaded via automated pipelines from other disease-related databases. Downloaded annotations are mapped to the same disease vocabulary used for manual annotations to provide consistency across the dataset. RGD also maintains disease-related quantitative phenotype data for the rat (PhenoMiner).

Online Mendelian Inheritance in Man (OMIM)

The Online Mendelian Inheritance in Man (OMIM) is a database that catalogues all the known diseases with a genetic component, and predicts their relationship to relevant genes in the human genome and provides references for further research and tools for genomic analysis of a catalogued gene. OMIM is a comprehensive, authoritative compendium of human genes and genetic phenotypes that is freely available and updated daily. The database has been used as a resource for predicting relevant information to inherited conditions.

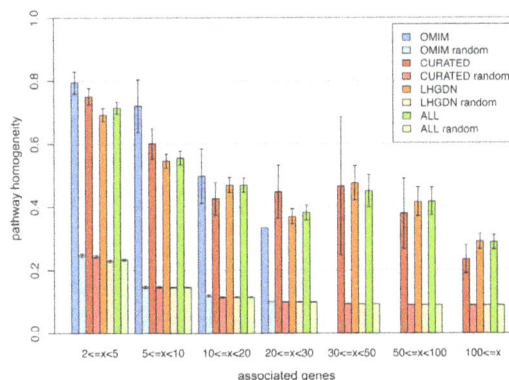

Pathway Hogeneity vs Associated Genes Showing the concept that diseases have large association with a variety of genes, a mean pathway homogeneity values of single diseases and random controls are plotted for four networks binned by the number of associated gene products per disease. The above graph shows how difficult is to correlate a bigger number of diseases vs concordance in 4 different databases, hence Gene Disease Databases test these relationships.

Ensembl SIFT Tool

This one of the largest resources available for all genomic and genetic studies, it provides a centralized resource for geneticists, molecular biologists and other researchers studying the genomes of our own species and other vertebrates and model disease organisms. Ensembl is one of several well-known genome browsers for the retrieval of genomic-disease information. Ensembl imports variation data from a variety of different sources, Ensembl predicts the effects of variants. For each variation that is mapped to the reference genome, each Ensembl transcript is identified that overlap the variation. Then it uses a rule-based approach to predict the effects that each allele of the variation may have on the transcript. The set of consequence terms, defined by the Sequence Ontology (SO) can be currently assigned to each combination of an allele and a transcript. Each allele of each variation may have a different effect in different transcripts. A variety of different tools are used to predict human mutations in the Ensembl database, one of the most widely used is SIFT, that predicts whether an amino acid substitution is likely to affect protein function based on sequence homology and the physic-chemical similarity between the alternate amino acids. The data provided for each amino acid substitution is a score and a qualitative prediction (either 'tolerated' or 'deleterious'). The score is the normalized probability that the amino acid change is tolerated so scores near 0 are more likely to be deleterious. The qualitative prediction is derived from this score such that substitutions with a score < 0.05 are called 'deleterious' and all others are called 'tolerated'. SIFT can be applied to naturally occurring nonsynonymous polymorphisms and laboratory-induced missense mutations, that will lead to build relationships in phenotype characteristics, proteomics and genomics

Literature Databases

This sort of databases summarize books, articles, book reviews, dissertations, and annotations about gene-disease databases. Some of the following are examples of this type: GAD, LGHDN and BeFree Data.

Genetic Association Database

The Genetic Association Database is an archive of human genetic association studies of complex diseases. GAD is primarily focused on archiving information on common complex human disease rather than rare Mendelian disorders as found in the OMIM. It includes curated summary data extracted from published papers in peer reviewed journals on candidate gene and genome Wide Association Studies (GWAS). The GAD was frozen as of 09/01/2014 but is still available for download.

Literature-derived Human Gene-disease Network

The literature-derived human gene-disease network (LHGDN) is a text mining derived database with focus on extracting and classifying gene-disease associations with respect to several biomolecular conditions. It uses a machine learning based algorithm to extract semantic gene-disease relations from a textual source of interest. It is part of the Linked Life Data, of the LMU in Munchen, Germany.

BeFree Data

Extracts gene-disease associations from MEDLINE abstract using the BeFree system. BeFree is composed of a biomedical Named Entity Recognition (BioNER) module to detect diseases and genes and a relation extraction module based on morphosyntactic information.

Integrative Databases

This sort of databases include Mendelian, compound and environmental diseases in an integrated gene-disease association archive and show that the concept of modularity applies for all of them They provide a functional analysis of diseases in case of important new biological insights, which might not be discovered when considering each of the gene-disease associations independently. Hence, they present a suitable framework for the study of how genetic and environmental factors, such as drugs, contribute to diseases. The best example for this sort of database is DisGeNET.

Gene Disease Associations Database DisGeNET

DisGeNET is a comprehensive gene-disease association database that integrates associations from several sources that covers different biomedical aspects of diseases. In particular, it is focused on the current knowledge of human genetic diseases including Mendelian, complex and environmental diseases. To assess the concept of modularity of human diseases, this database performs a systematic study of the emergent properties of human gene-disease networks by means of network topology and functional annotation analysis. The results indicate a highly shared genetic origin of human diseases and show that for most diseases, including Mendelian, complex and environmental diseases, functional modules exist. Moreover, a core set of biological pathways is found to be associated with most human diseases. Obtaining similar results when studying clusters of diseases, the findings in this database suggest that related diseases might arise due to dysfunction of common biological processes in the cell. The network analysis of this integrated database points out that data integration is needed to obtain a comprehensive view of the genetic landscape of human diseases and that the genetic origin of complex diseases is much more common than expected.

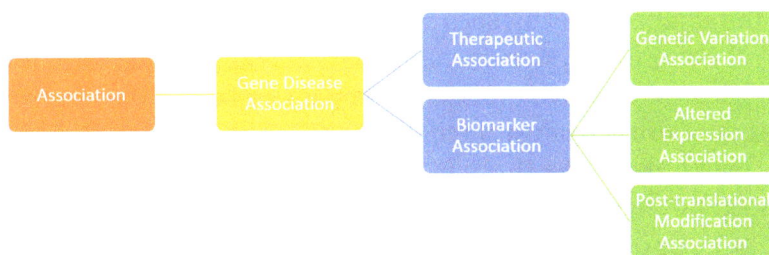

DisGeNET Gene-disease Association Ontology

The description of each association type in this ontology is:

- Therapeutic Association: The gene/protein has a therapeutic role in the amelioration of the disease.

- Biomarker Association: The gene/protein either plays a role in the etiology of the disease (e.g. participates in the molecular mechanism that leads to disease) or is a biomarker for a disease.

- Genetic Variation Association: Used when a sequence variation (a mutation, a SNP) is associated to the disease phenotype, but there is still no evidence to say that the variation causes the disease. In some cases the presence of the variants increase the susceptibility to the disease. In general, the NCBI SNP identifiers are provided.

- Altered Expression Association: Alterations in the function of the protein by means of altered expression of the gene are associated with the disease phenotype.

- Post-translational Modification Association: Alterations in the function of the protein by means of post-translational modifications (methylation or phosphorylation of the protein) are associated with the disease phenotype.

Future in Gene Disease Databases

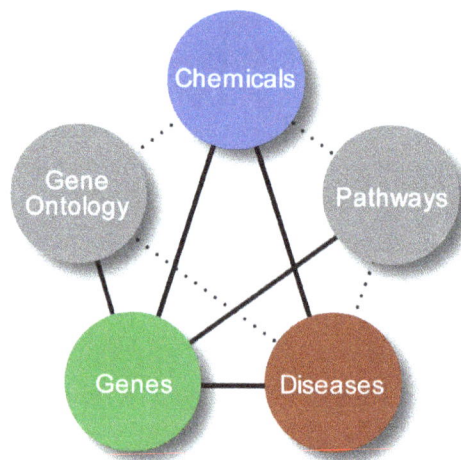

Relationships in Gene Diseases

The completion of the human genome has changed the way the search for disease genes is performed. In the past, the approach was to focus on one or a few genes at a time. Now, projects like the DisGeNET exemplify the efforts to systematically analyze all the gene alterations involved in a single or multiple diseases. The next step is to produce a complete picture of the mechanistic aspects of the diseases and the design of drugs against them. For that, a combination of two approaches will be needed: a systematic search and in-depth study of each gene. The future of the field will be defined by new techniques to integrate large bodies of data from different sources and to incorporate functional information into the analysis of large-scale data generated by bioinformatics studies.

Bioinformatics is both a term for the body of biological gene disease studies that use computer programming as part of their methodology, as well as a reference to specific analysis pipelines that are repeatedly used, particularly in the fields of genetics and genomics. Common uses of bioinformatics include the identification of candidate genes and nucleotides, SNPs. Often, such identification is made with the aim of better understanding the genetic basis of disease, unique adaptations, desirable properties, or differences between populations. In a less formal way, bioinformatics also tries to understand the organisational principles within nucleic acid and protein sequences.

The response of bioinformatics to new experimental techniques brings a new perspective into the analysis of the experimental data, as demonstrated by the advances in the analysis of information from gene disease databases and other technologies. It is expected that this trend will continue with novel approaches to respond to new techniques, such as next-generation sequencing technologies. For instance, the availability of large numbers of individual human genomes will promote the development of computational analyses of rare variants, including the statistical mining of their relations to lifestyles, drug interactions and other factors. Biomedical research will also be driven by our ability to efficiently mine the large body of existing and continuously generated biomedical data. Text-mining techniques, in particular, when combined with other molecular data, can provide information about gene mutations and interactions and will become crucial to stay ahead of the exponential growth of data generated in biomedical research. Another field that is benefiting from the advances in mining and integration of molecular, clinical and drug analysis is pharmacogenomics. *In silico* studies of the relationships between human variations and their effect on diseases will be key to the development of personalized medicine. In summary, Gene Disease Databases have already transformed the search for disease genes and has the potential to become a crucial component of other areas of medical research.

Model Organism Databases

Principles of Model Organism Databases

Recent advances in DNA sequencing technologies over the past two decades have led to an increase in the number of fully sequenced genomes and other types of publicly available DNA sequences, which has in turn allowed a great expansion in the depth and breadth of experimental data available to today's researcher. In order to make the most of this information, it must be collected, vetted, collated, and made available to the relevant scientific community (i.e., it must be curated). This curation occurs within the context of Model Organism Databases (MODs), which are assuming increasing importance in all areas of biology.

"Model organisms" are nonhuman organisms that are typically used for biological research. The resulting data can be used as a framework for the interpretation and understanding of similar data from humans or other medically or economically important species. Popular model organisms include budding yeast, fruit flies, and laboratory mice, all of which contain genes that encode proteins and other gene products similar to those found in humans. Genetic manipulation of model organisms is generally the most efficient path to understanding the effects of mutations in their human homologs. Model organisms have become especially effective reference species because vast amounts of data have been generated, collected, and made freely available to the public research community.

MOD Functions

All MODs serve a variety of functions, the most important among them being the organization and presentation of experimental data from disparate sources. Think of any particular MOD as the central hub of that organism's research community; they are designed to clearly and concisely present research regarding a key organism to all biologists, regardless of specialty. The strength of MODs lies in the fact that the data contained in them are meticulously curated from the primary literature by experts, thereby providing centralized, impartial summaries of various types of biological information for use by researchers. When organized well, the juxtaposition of different types of information within a MOD presents researchers with an expanded view of the roles of the genes and gene products within a cell, thereby facilitating the formulation and testing of new hypotheses. For example, showing on a single Web page that a gene is expressed when a new sugar is introduced into the growth medium and contains a DNA-binding domain may help a researcher infer that this gene encodes a transcription factor that activates genes needed to digest the sugar. Types of data typically presented at MODs are genomic sequence and mapping data, gene expression patterns and functional characterizations, homology data, mutant phenotypes, allele variants, quantitative trait loci (QTLs), biochemical pathways, protein structures, and historical nomenclatures, as well as the primary literature from which all of this information is derived. However, the exact kinds of data presented at a particular MOD depend entirely on the experiments researchers have performed using that organism.

MODs exist as service organizations rather than research organizations. The primary function of the scientific curators, the biological experts employed by the MOD, is not to perform experiments, but rather to facilitate the open exchange of scientific information.

As such, they do not produce the data displayed by the MOD; instead, they obtain and present data from peer-reviewed journals, referencing the information to ensure validity and accountability. This occurs through the unbiased, standardized presentations of data and maintenance of close relationships with the communities they serve and with staff at other MODs.

Gene Ontology

Most MODs foster relationships with other databases to share data, develop annotation tools, and ensure consistency of the biological annotation of homologs across species. The Gene Ontology (GO) is a well-known and useful product of these interactions. The GO Consortium is a collaborative effort composed of several MOD groups and other bioinformatics groups who have come together to develop controlled vocabularies for the annotation of gene products in a wide variety of organisms.

These controlled vocabularies, known as ontologies, consist of standardized terms (i.e., kinase activity, transsulfuration, mitochondrion, etc.) with controlled definitions, and include all known relationships between the terms (a "histone kinase" is a type of "protein kinase;" a "protein kinase" is a type of "phosphotransferase" etc.). Since ontologies are collectively defined and maintained by the participating MODs, using terms in the ontology to describe biological entities in all species guarantees that the language used will be consistent across research groups and scientific communities. This uniformity in representing and communicating biological knowledge improves inferences that can be made from experimental data, simplifies computational searches, and allows users to find similar data and types of information in different MODs.

The GO ontologies are divided into three domains that are needed for gene annotation in all organisms: Molecular Function, Biological Process, and Cellular Component. Molecular Function refers to the tasks or activities performed by individual gene products, such as transcription factor, lyase activity, or electron carrier, etc. Biological Process describes broad biological series of reactions, such as mitosis, purine metabolism, or membrane docking, etc. Cellular Component encompasses subcellular locations, structures, ormacromolecular complexes, such as nucleus, microtubule, or origin recognition complex, etc. The three ontologies together contain >20,000 terms. Terms in the structured vocabularies are used for the annotation of gene products (proteins or RNAs) based on published experimental evidence. The annotations made by MOD curators are incorporated into their own databases, and are provided to the GO Consortium for dissemination through its. For example, Ono et al. characterize the *CYS4* gene in *S. cerevisiae*, which codes for cystathionine beta-synthase, and provide evidence that this activity is involved in the biosynthesis of cysteine via a cystathionine intermediate.

MOD Tools

The content of active MODs is constantly being updated and expanded, both through the curation of newly published information and through the development of new data analysis tools and visualization interfaces. The main point of entry for most MODs is the home page, and the basic unit of organization typically focuses on individual genes. Users can perform basic searches using gene names or keywords, or more complex queries of various types of data using specially designed search interfaces. Data can also be analyzed using various Web-based applications, or downloaded in bulk via interfaces or FTP (File Transfer Protocol). A MOD will often provide a site map and online help documentation describing various aspects and available tools, as well as direct help for users via e-mail interaction with the MOD's scientific curators.

This unit provides basic protocols for accessing information about genes in MODs. The growing number of MODs and the various types of data and analysis tools available from them cannot all be covered in this unit. The aim of this set of protocols is to provide a general introduction to enable the novice user to gain entry into various characteristic MODs, then find and retrieve basic information about genes. The unit will explain simple uses for two tools found at many MODs, GBrowse and Textpresso, both of which were designed as part of the Generic Model Organism Database (GMOD) project. GMOD began as a collaboration between four established genome databases—SGD, FlyBase, MGD, and WormBase—to develop and provide generic database architecture and software to the scientific community under an open source policy. The goal of the GMOD project is to generate a set of independent software components that can be mixed and matched to set up MODs for newly sequenced genomes in an efficient and cost-effective manner, without unnecessary duplication of effort in the development of curation and visualization software. The result is that many MODs share common components, making it easier and more intuitive for users to navigate the different layouts at diverse MODs. GBrowse is the genome feature browser Web application produced by GMOD, whereas Textpresso is the full-text literature search Web application.

Cambridge Structural Database

The Cambridge Structural Database (CSD) contains a complete record of all published organic and metal–organic small-molecule crystal structures. The database has been in operation for over 50 years and continues to be the primary means of sharing structural chemistry data and knowledge across disciplines. As well as structures that are made public to support scientific articles, it includes many structures published directly as CSD Communications. All structures are processed both computationally and by expert structural chemistry editors prior to entering the database. A key component of this processing is the reliable association of the chemical identity of the structure studied with the experimental data. This important step helps ensure that data is widely discoverable and readily reusable. Content is further enriched through selective inclusion of additional experimental data. Entries are available to anyone through free CSD community web services. Linking services developed and maintained by the CCDC, combined with the use of standard identifiers, facilitate discovery from other resources. Data can also be accessed through CCDC and third party software applications and through an application programming interface.

Value of Sharing Crystal Structures

The ongoing stewardship of the Cambridge Structural Database (CSD) has been the core activity of the Cambridge Crystallographic Data Centre (CCDC) since its inception in 1965. The CCDC is committed to providing a permanent archive of crystal structures and making these available to all. This non-profit, charitable organization is overseen by an international board of trustees drawn from the community it serves.

The CSD contains all published organic and metal–organic small-molecule crystal structures whose structures have been determined using the technique of crystallography. In addition, it acts as a publication vehicle for structure determinations with no accompanying manuscript.

Specifically, the CSD contains both X-ray and neutron diffraction analyses from a single-crystal study or a powder study where cell parameters, atomic coordinates and refinement are reported. To ensure comprehensive coverage of single-crystal data, cell parameters and all available data are included even if no coordinates are available. Powder structures without coordinates are available from the International Centre for Diffraction Data .

The CSD covers all organic and metal–organic structures, where organic is generally taken to mean a carbon-containing molecule. The CSD also contains boron compounds containing one or more B—H or B—OH bond and borazines and ring compounds containing any two of the following elements: N, P, S, Se and Te. Purely inorganic structures that do not fit the criteria above are added to the Inorganic Crystal structure Database produced by FIZ Karlsruhe or the Metals Database. for metals and alloys. Peptides and polysaccharides of up to 24 residues and mono-, di- and tri-nucleotides are included in the CSD, higher oligomers are covered by the Nucleic Acids Database with the Protein Data Bank curating and sharing structural data of larger biological macromolecules. In all cases, these guidelines are relaxed where there is clear scientific merit in including a structure in multiple resources.

The database provides value in two distinct ways. The first simply relates to the aggregation and standardization of structures, which facilitates access to individual entries. This brings value both to the data generators and consumers. A single archive of all structures allows crystallographers to avoid the inadvertent redetermination of structures and provides a mechanism by which they can archive the output of their work at a specialist data centre for their own future use. Of course, it also allows for the easy sharing of their work, massively increasing their sphere of influence. Such sharing has always been the norm for the crystallographic community and the use of this worldwide, standard, specialist discipline repository allows individuals to demonstrate their adherence to the new data sharing mandates of various funding bodies. As all entries are subject to both automatic and manual curation, they can usually be used without further processing. Indeed, one might argue that the financial cost of maintaining such a resource, although significant, is recovered many times over by removing the need for repeated correction by users.

A further vital property of the CSD is its comprehensive and up-to-date nature. As it represents the complete record of published structures and is updated within a few moments of a new publication, users can have confidence that there are no published structures of relevance of which they are unaware.

The second distinct benefit of the database comes from the study of the collection of entries. This was perhaps best articulated by the founder of the CSD, Dr Olga Kennard, who, recounting a discussion with JD Bernal commented, 'We had a passionate belief that the collective use of data would lead to the discovery of new knowledge which transcends the results of individual experiments'. The motivations behind the determination of crystal structures do differ, the most common probably being the confirmation of a molecule's chemical identity. However, the use they are put to once in the database usually bears no similarity to these motivations. Two illustrative examples are perhaps the establishment of the ability of C—H groups to act as hydrogen-bond donors and 'structure correlation' – the linking of three-dimensional geometry to reaction pathways.

This 'new knowledge' relates primarily to the geometry of molecules and the interactions they make. A knowledge of these factors underpins huge areas of both fundamental and applied science. They form the basis of our understanding of the energetics of molecular conformation – from bond lengths and angles, through to torsional preferences. They also teach us about the fundamentals of molecular recognition, be it small molecules interacting with small molecules in a lattice or with a protein.

Development of the CSD over a 10-year timespan from 2006 to 2015

	2006	2015
Number of CSD entries	400 374	800 239
Number of compounds	363 372	731 675
Number of associated articles	232 858	408 899
New entries	34 030	60 122
Entries classed 'Organic'	43%	43%
Entries with R-factor < 10%	92%	94%
Average atoms per structure	68.6	80.6
Polymeric entries	7%	11%

Progression in CSD

In 2015 the number of entries in the CSD surpassed 800 000. This is twice the number of entries in the database less than a decade ago. Comparing statistics based on the database as it was then allows us to see what has changed in the last decade – and what has not. Table above shows that the proportion of structures which are organic or metal–organic structures (which we classify as structures containing a transition metal, lanthanide, actinide, or Al, Ga, In, Tl, Ge, Sn, Pb, Sb, Bi, Po) has remained fairly constant. What has changed is the complexity of the structures being published: the average number of atoms per structure and the average molecular weight have increased, as has the proportion of structures that are polymeric or that have resolved disorder.

Publication sources for CSD entries.

Journal (Publisher)	% CSD
Inorg. Chem. (ACS)	8
Dalton Trans. (RSC)	6
Organometallics (ACS)	6
J. Am. Chem. Soc. (ACS)	5
Acta Cryst. Section E (IUCr)	5
J. Organomet. Chem. (Elsevier)	3
Chem. Commun. (RSC)	3
Acta Cryst. Section C (IUCr)	3
Inorg. Chim. Acta (Elsevier)	3
Chem. Eur. J. (Wiley)	3
Polyhedron (Elsevier)	3
Angew. Chem. Int. Ed. (Wiley)	3
Eur. J. Inorg. Chem. (Wiley)	2
J. Org. Chem. (ACS)	2
CrystEngComm (RSC)	2
Cryst. Growth Des. (ACS)	2
Acta Cryst. Section B (IUCr)	2
CSD Communications (CCDC)	2
Z. Anorg. Allg. Chem. (Wiley)	2
Tetrahedron (Elsevier)	2

Another significant change is in the number of new structures published per year, which in 2015 was almost twice the number published during 2006. Structures in the CSD are associated with over 400 000 articles from 1600 publication sources. Table above shows the top 20 publication sources currently represented in the CSD; these account for 67% of the entries in the database.

The increase in volume and complexity of structures deposited into the CSD over the past decade has presented both administrative and scientific challenges.

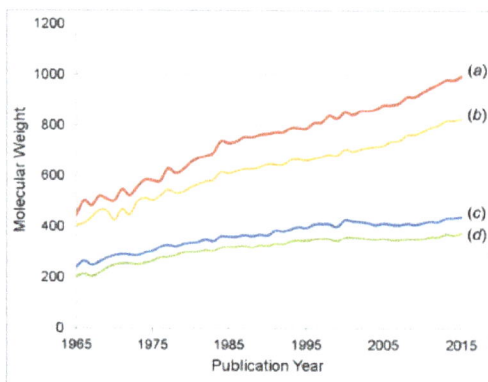

The increase in the average molecular weight of CSD entries since 1965, with (a) average formula weight per Z' of metal–organic structures, (b) average molecular weight of heaviest component of metal–organic structures, (c) average formula weight per Z' of organic structures and (d) average molecular weight of the heaviest component of organic structures.

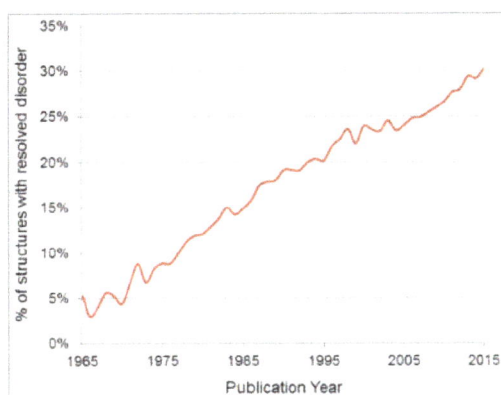

The increase in resolved disorder in CSD entries since 1965.

Deposition and Retrieval of Data

The method for deposition into the CSD has evolved since the advent of the CIF format in the 1990s, when email depositions dominated. In 2009 the CCDC launched an online web-based tool which is now the main route for deposition. In 2015, 90% of structures were deposited with the CCDC prior to publi- cation and 85% were submitted through this service. A key benefit of this early deposition is that at this point the crys- tallographer who generated the data is likely to be the depositor and be in a position both to provide the richest data and to respond to any issues most effectively.

During deposition, the CIF syntax is automatically checked based on the checks in enCIFer and the depositor is required to fix any issues before continuing with the deposition process. Depositors are strongly encouraged to deposit structure factors in line with the IUCr's publication standards for crystal structures. The embedding of reflection data into the CIF by structure refinement programs such as SHELXL greatly simplifies this process. As a consequence the amount of reflection data stored at CCDC has increased significantly and the majority of new depositions now include this data.

In 2015 the checkCIF/PLATON service was integrated into the CCDC's deposition process. This allows the researcher to generate validation reports and embed validation responses into the 'structure of record' during deposition to the CSD.

During deposition, key metadata are extracted from the CIF and presented to the researcher giving them the opportunity to check and further enhance the data that is shared through the CSD. To aid with this process a three-dimensional representation of the structure is displayed to the depositor using JSmol.

On deposition, each dataset is assigned an accession identifier referred to as a 'CCDC number' in the format CCDC 1234567 (older entries have six digits). This uniquely identifies the data associated with a particular structure determination and persists for the lifetime of the dataset. CCDC numbers are communicated to the depositor once it has been confirmed that the dataset is not a duplicate submission, usually within seconds of deposition. Should the data require further investigation, CCDC Deposition Coordinators address any problems, minimizing the delay in providing the CCDC number to the depositor. A reference code (known as a CSD refcode and in the format

ABCDEF) is assigned to structures as they are indexed into the CSD itself. Where possible, determinations of the same substance are assigned into a CSD refcode family (with a CSD refcode format of ABCDEF01). These codes are used as a common way of referring to structures extracted from the CSD-System.

CCDC numbers are used in manuscripts to indicate the location of the data that supports results described in that article. They are used as the basis for providing links directly to the structure from within the article when published. Prior to publication, data is stored in a confidential data archive and is only available to referees and publishers during the peer review process. Providing access to the structure of record, reflection data and validation reports helps ensure the accuracy and integrity of published science. Authors and depositors are able to revise data stored at CCDC up to the point of publication and retain a consistent CCDC number. About one quarter of all submissions are revised during the deposition process. In 2015 alone, over 67 465 unique CCDC numbers were assigned and data was deposited by over 10 000 unique depositors.

For structures associated with a journal article, computational workflows and processes with the major publishers handle the flow of data during the publication process. The publication of an article referencing a dataset results in the immediate public release of the corresponding structure. This is automatically triggered by feeds from the publisher or, failing that, by the identification of a manuscript with a CCDC number. Structures deposited with the intention of accompanying a publication are held securely in trust for a period of 1 year. If no publication is identified within that year, authors are contacted to confirm that the structure should be published as a CSD Communication. If the structure is still intended for another publication then the embargo period is extended for another year.

The validation step of the CSD Deposition process showing the integration with checkCIF.

A structure published as a CSD Communication is freely available through the CSD within seconds of CCDC number assignment. As the general appreciation of the value of data sharing has increased so has the popularity of publishing structures directly through the CSD. In fact, although only 2% of structures in the CSD are CSD Communications, 2015 saw over 4400 CSD Communications published, making it likely that this will soon become the most popular way in which to publish crystal structures.

At the point of publication, entries are available to anyone through free CSD-Community web services. These services allow anyone to access published structures either via the CCDC website or by following links from other resources.

They provide an interactive visualization of the three-dimensional structure through JSmol (Hanson et al., 2013), a two-dimensional chemical diagram and key metadata associated with the entry. Individual data files can also be downloaded and used by anyone wishing to investigate or explore the entries in more detail.

CSD community Access Structures results page. This is the 'landing page' for many referring services based, for example on DOI or CCDC number.

Creation of the CSD

Processing Entries

Due to the rising number of structures, depositions and transactions a new processing system, CSD-Xpedite, has been developed, automating most informatics processes required to manage depositions and process crystal structures into entries in the CSD. Figure shows how data flows through the system and where various users interact with the system.

CSD-Xpedite is built around Microsoft Dynamics CRM and SharePoint. The system is designed to store all data in one unified system and many automated processes have been incorporated to reduce the number of manual interventions required to process entries through the CSD creation process. This scalable system allows for the ever increasing rate of deposition, and its modular nature allows extra functionality to be incorporated with minimal disruption. One important aspect of CSD- Xpedite is the use of a new extensible file format for internal storage of CSD entries. This format allows a fuller representation of the underlying data than in previous formats (BCCAB, ASER). Databases for use in the CSD-System are created from this master database.

A key component in CSD-Xpedite is CSD-Editor, an interactive tool for processing structures such that they can enter into the CSD. This uses many of the visualization and menu options available in the CSD-System program Mercury. Deposited CIFs can be easily viewed alongside any associated article. Errors and warnings are displayed for each structure to allow expert structural chemistry editors to concentrate their efforts on the challenging scientific parts of the process.

Assignment of Chemical Identity

An important aspect of creating the CSD is the assignment of the chemical representation to structures so that these can be reliably searched and analysed using structure-based methodologies such as substructure search. In the past, the chemical representation was assigned by scientific editors visualizing the structure and consulting any associated article. A program called decipher. Now helps automatically assign 'chemistry' to structures.

DeCIFer uses the information already in the CSD to interpret a new structure and add a chemical representation to the atomic coordinates in the CIF. The stages involved in the automatic creation of the CSD entry include resolution of any disorder in the structure, detection of bonds and de-termination of bond types and charges. Assignment of bond types, charges and the inference of missing H atoms uses a probabilistic Bayesian approach which allows all the existing entries in the CSD to be used to help assign the chemistry to new entries.

The final step of the deCIFer process is to validate the assignment by looking at improbable fea-tures such as unprecedented bond types, unlikely oxidation states, unlikely metal– metal bonds and other empirical indicators such as non-planar double bonds. From this analysis a reliability score is calculated which highlights any possible errors in the representation of the structure which can then be reviewed and corrected during CSD entry validation. It is important to note that this reliability score simply gives a confidence value to the automatic treatment of the entry; this is not necessarily related to the 'quality' of the structure.

With 'chemistry' assigned, an internally developed diagram generation algorithm takes the three-dimensional coordinates of an entry and 'flattens' these, producing a two-dimensional rep-resentation with minimal overlaps of atoms and bonds. This procedure also takes advantage of the many two-dimensional diagrams already available for related entries, often drawn previously by scientific editors, which increases both the quality and consistency of diagrams. Although the au-tomatic decipher process is used as much as possible structures are still manually viewed by expert structural chemists before they are added to the CSD.

Using the CSD

Identifying and Linking Digital Objects

At the point a structure can be made public a Digital Object Identifier (DOI) is associated with the deposited dataset (580 000 are currently available). This DOI allows a third party to link to a structure summary page without needing to know details of linking services provided by the CCDC. In addition the DOI provides the basis for a more formal citation of the deposited dataset in line with the spirit of the Joint Declaration of Data Citation Principles. The metadata provided to the DOI registration agency DataCite indicates key elements of a dataset citation including contributor, a title and publication year. This enables citation of the dataset independently of the associated article and thus helps ensure recognition of the specific contribution made by the crystallographer, who may not always be included in the article's author list.

Metadata submitted by the CCDC to DataCite when generating a DOI is openly accessible and facilitates inter- operability with third party systems to improve the discover- ability of data. Examples where other parties have taken advantage of this include the Thomson Reuters Data Citation Index, where data is currently available for 530 000 CSD entries, and the prototype RDA/ WDS Data-Literature Interlinking service. In order to support this interoperability, additional metadata items are made public including the DOIs of associated articles and the chemical name of the substance studied. Discoverability of data by chemists and biologists is enabled by establishing links to datasets from services such as Chem Spider and PubChem. The overlap between these resources and the CSD is identified by taking advantage of the International Chemical Identifier Standard (InChI) which provides a unique and canonical representation of the chemical substance studied. Links from ChemSpider and PubChem have been established for over 52 000 compounds that could be reliably identified using InChIs as being in common between these resources and the CSD. InChIs have also been used to identify correspondences between CSD entries and ligands bound to macromolecules in structures archived in the Protein Data Bank. This lookup is enabled by a CCDC web service that identifies the best representative CSD entry for a given molecule and provides access to its coordinates. This is a particularly useful resource for structural biologists refining or investigating the structure of a protein ligand. Representative structures are freely available for approximately 1500 PDB ligands in a Chemical Component Model file provided by the PDB.

Accessibility and Efficiency

The CSD community web services provide free access to the entire collection of structures. As well as these services there are a number of other avenues to explore and exploit the data ranging from free lookup tools such as CellCheckCSD to advanced search, analysis and validation tools in the CSD-System. More specialist applications are provided as part of the CSD-Enterprise suite.

CellCheckCSD is an automated tool which uses the data from the CSD to check unit cells during data collection and can be used to check that the structure is novel rather than the starting material, a by-product or another previously determined structure.

The CSD-System enables knowledge to be gained from the collection of data through powerful two-dimensional/three dimensional search capabilities, extensive geometry analysis tools, inter- and intramolecular interaction analysis, generation of high impact graphics and the ability to delve further into the data using a Python-based applications programming interface.

Visualizing three-dimensional structural data can be a powerful way to teach chemistry concepts. Through the CSD community services, educators worldwide are able to access all 800 000 entries in the CSD, but the enormity of the database means it is not a simple process to identify the key structures that are most appropriate for the class. To this end a CSD teaching database has been compiled. This is a collection of around 700 carefully selected crystal structures, targeted to represent a diverse range of chemistry and to allow teachers to demonstrate key chemical concepts and principles.

To support the free archiving services and free universal access to all individual structures, users of client installed software (i.e. the CSD-System) are asked to contribute to the running costs of the CCDC. In many cases this is done at a national level, allowing unfettered use, with the modest contribution levels set according to their economic status. These contributions fund the maintenance of the CSD, whereas larger contributions from for-profit users of the system fund its ongoing development.

Saccharomyces Genome Database

The *Saccharomyces* Genome Database (SGD) collects and organizes information about the molecular biology and genetics of the yeast *Saccharomyces cerevisiae*. The latest protein structure and comparison tools available at SGD are presented here. With the completion of the yeast sequence and the *Caenorhabditis elegans* sequence soon to follow, comparison of proteins from complete eukaryotic proteomes will be an extremely powerful way to learn more about a particular protein's structure, its function, and its relationships with other proteins

The *Saccharomyces* Genome Database (SGD) exists to provide the scientific community with access to the *Saccharomyces cerevisiae* sequence and the wealth of associated information. This database includes a variety of biological information, including the complete, annotated DNA and protein sequence along with several tools for sequence analysis. Here we focus on features of SGD that provide users with tools for comparing yeast protein sequences and examining protein structure. Sequence comparisons play a critical role in the initial process of determining the function of specific proteins and also in interpreting new protein sequence data from large-scale genome sequencing projects. There are several sequence comparison tools at SGD. Here, we discuss the Genome-wide Protein Similarity View program, which is a powerful tool for examining protein similarities. Like the expanding base of sequence information, there is also a growing amount of structural information. Sacch3D is a feature of SGD that organizes and presents structural information about yeast proteins and their putative homologs. Familiarity with tools will enable molecular biologists and geneticists to gain insight into the function and possible evolution of their protein of interest.

Examining Protein Similarities at SGD

The Genome-wide Protein Similarity View (GPSV), displays, either graphically or in a, all the ORFs in the *S.cerevisiae* genome that are similar to a given query ORF based on a Smith-Waterman protein sequence comparison. Smith-Waterman comparisons were conducted on a TimeLogic De-Cypher II machine using the affine Smith-Waterman application. This system uses the pscorer program to calculate a *P*-value.

The GPSV graphic view, a Java applet, represents all 16 yeast nuclear chromosomes as horizontal black bars, with centromeres and positional coordinates indicated. Superimposed on the black chromosome bars are small vertical colored bars (similarity bars) that represent ORFs predicted by the Smith-Waterman analysis to have significant protein sequence similarities. A small black rectangle surrounds the bar for the query ORF itself. Its ORF name and associated standard gene name are displayed in the upper right hand corner. The color of the bars indicates the relative similarity shared with the query ORF. The warm colors (red) indicate high similarity while the cool colors (blue) indicate lower similarity. The user can switch between different query ORFs, add ORFs to the query list, and change several parameters of the similarity display.

Immediately below the graphic display are seven fields that contain additional information about the query and target ORFs. The first field displays a constantly updated readout of the current location of the mouse cursor in terms of base pairs along the chromosomes and the names of genes or ORFs selected by the mouse. The remaining fields contain information when the cursor is positioned over a similarity bar. The classes of information are: (i) P-Value (the P-value for the similarity between the query ORF and the target ORF); (ii) % Aligned (the percent of the query sequence that is aligned with the target sequence); and (iii) Gaps (the number of gaps inserted in the query sequence to achieve the alignment).

Each similarity bar can be clicked to reach more information about the target ORF. Options include links to the SGD Locus and Gene/Sequence Resources pages for the target ORF, an alignment of the query and target amino acid sequences, the DNA Similarity View, which displays the alignment of the target and query ORF DNA sequences, and the Protein Similarity View, where the selected target ORF is used as the query ORF.

The protein similarity data can also be displayed as a table, which can be accessed from the graphic display page or from the ORF input form. The table lists target ORFs in order of decreasing similarity to the query ORF, as determined by P-value; the target ORFs can also be sorted by percent identity. For each target ORF, the table lists the same information and links as the graphic display.

Protein Structure at SGD

The Sacch3D feature provides structural information for *S.cerevisiae* proteins by integrating data from SGD and structural databases and presenting it via up-to-date, concise summaries and links to structural resources. Sacch3D supplies researchers both within and outside the yeast community with insight into the structure and putative function of yeast proteins. Structural information for Sacch3D is obtained primarily by BLASTP analysis of the Brookhaven Protein Database (PDB) to identify all PDB structures with significant sequence (and therefore likely structural) similarity to yeast proteins. Results are updated monthly to keep pace with the growth of the PDB. To reduce the redundancy in the PDB and thus simplify the BLAST analysis, all PDB protein sequences are first clustered into groups of closely related sequences before the BLAST is run. As of September 1998, 18% of yeast proteins have either a known structure or putative homolog in a clustered version of the PDB. The Sacch3D search utility provides a structural information page for all ORFs in the yeast genome. This page contains information provided by both internal and external resources. A summary table is presented showing PDB structures for the yeast protein (if a structure can be identified) and proteins with which it shares significant sequence similarity. For each structure, there are links to a variety of freely available 3D viewers and external structural databases. 3D viewers include RasMol, Webmol Java viewer, Chime (MDL Information Systems) and Cn3D. External structural databases include PDB, SCOP, CATH, PDBsum, ModBase, Macromolecular Movements Database and MMDB. The PDB similarities are listed from best to worst (based on

BLASTP *P*-value) and are clustered to facilitate browsing. That is, one representative structure is listed in cases where there are multiple variants of the same structure (mutants or complex forms). Access to the neighboring structures is also provided.

For yeast proteins without a known structure but with significant sequence similarity to proteins with structures contained within PDB, links are available on the structural information page for homology-based models of the yeast protein structure. These models are accessed by links to the external resources ModBase and Swiss-Model. Even for yeast proteins that lack significant similarities in the PDB, a variety of useful links are presented. These include links to secondary structure predictions, several pre-computed BLAST reports, and the Emotif and Pfam sequence search programs. Links to Swiss-Prot, Entrez and the NCBI COGs site for the yeast protein are also included.

Other Sacch3D features include: (i) flexible search options using yeast gene or ORF name, PDB identifier, Swiss-Prot identifier/accession, or text; (ii) a special page devoted to *S.cerevisiae* structures in PDB showing the number of different structures for each yeast protein with links to SGD and Sacch3D; (iii) a structural URLs page. Sacch3D maintains this list of URLs for web sites relevant to the analysis of protein structure and/or function, including links to structural biology resources, 3D viewers, genome-analysis web sites, journals and research groups; (iv) a domains page providing access to yeast proteins based on their SCOP-classified domains. Users can search for domains using a yeast gene/ORF name, a SCOP class number, or a SCOP fold number. Links are also provided to the WebMol Java viewer to illustrate the location of the domain within the context of the 3D structure; (v) an analysis page that performs an electronic version of a Southern blot using the yeast genomic sequence; (vi) a What's New page that lists new features in Sacch3D as well as new yeast protein structures and new protein structures with homology to yeast proteins.

References

- Wren JD, Bateman A (2008). "Databases, data tombs and dust in the wind". Bioinformatics. 24 (19): 2127–8. doi:10.1093/bioinformatics/btn464. PMID 18819940

- What-is-bioinformatics: scq.ubc.ca, Retrieved 09 April 2018

- Hubbard T, et al. (January 2002). "The Ensembl genome database project". Nucleic Acids Research. 30 (1): 38–41. doi:10.1093/nar/30.1.38. PMC 99161 . PMID 11752248. Retrieved 11 November 2014

- Bioinformatics: britannica.com, Retrieved 29 June 2018

- Davis, A.; Wiegers, T. (2013). "Text Mining Effectively Scores and Ranks the Literature for Improving Chemical-Gene-Disease Curation at the Comparative Toxicogenomics Database". PLoS ONE. 8 (4): 1–29. Bibcode:2013PLoSO...858201D. doi:10.1371/journal.pone.0058201

- Biological-databases-an-overview-and-future-perspectives: enago.com, Retrieved 29 March 2018

- Davis, A.; King, B. (2011). "The Comparative Toxicogenomics Database: update 2011". Nucleic Acids Res. 39 (1): 1067–1072. doi:10.1093/nar/gkq813

- Gene-disease-specific-databases: humanvariomeproject.org, Retrieved 16 May 2018

Sequence Alignment and Formats

In bioinformatics, the sequence of DNA, RNA and protein are arranged for identifying regions of similarity, based on structural, functional or evolutionary relationships. This process is termed as sequence alignment. The aim of this chapter is to explore the crucial aspects of biological sequence alignment, alignment-free sequence analysis and biological sequence format.

Biological Sequence Alignment

Biological sequence alignment is a widely used operation in the field bioinformatics and computational biology. It aims to find out whether two or more biological sequences (e.g., DNA, RNA, or Protein sequences) are related or not. This has many important real world applications. For instance, if some information about one of the sequences is already known (e.g., the sequence represents a cancerous gene) then this information could be transferred to the other unknown sequences, which could be vital in early disease diagnosis and drug engineering. Other applications include the study of evolutionary development and the history of species and their groupings.

As individual laboratories exchange more annotated biological data through comprehensive databases such as NCBI's retrieval system, Entrez, (which integrates GenBank1), researchers have recently become interested in detecting remote homologies by querying a sequence of interest against a subfamily of a distant lineage. In order to unveil the structural or functional importance of an unknown sequence, one conducts, as an initial procedure, a sequence alignment in the framework of the comparative computational biology. A sequence alignment is a way of arranging the primary sequences of DNA, RNA, or protein to identify the regions of similarity that may be a consequence of functional, structural, or evolutionary relationships between the sequences. The resulting alignment yields an edit transcript of mismatches and indels, i.e., insertions and deletions, where mismatches can be interpreted as point mutations and gaps as indels. As a result, we can infer how sequences with the same origin would diverge from one another.

DNA Alignment

Sequence comparison lies at the heart of the bioinformatics analysis. As new biological sequences are being generated at exponential rates, sequence comparison is becoming increasingly important to draw functional and evolutionary inference of new protein with proteins already existing in database. Sequence alignment is the process by which sequences are compared by searching for common character patterns and establishing residue-residue correspondence among related sequences.

The rapid evolution of sequencing techniques combined with the intense growth in the number of large-scale genome projects is producing a huge amount of biological sequence data. Nevertheless, determining the genome sequence is only the first step toward deciphering the genetic message encoded in those sequences. In genome projects, newly determined sequences are first compared with those placed in genomic databases, in order to discover similarities. This is done because relevant sequence similarity is evidence of common evolutionary origin and homology relationship. Sequence comparison is, therefore, a very basic but important step in genome projects. As a result of this step, one or more sequence alignments can be produced. A sequence alignment has a similarity score associated to it that is obtained by placing one sequence above the other, making clear the correspondence between the characters.

Methods of Sequence Alignment

Global Alignment

Two sequences to be aligned are assumed to be generally similar over their entire length. Alignment is carried out from beginning to the end of both the sequences to find the best possible alignment.

Local Alignment

This method of alignment does not assume that the two sequences have the similarity over the entire length. It only finds local regions with the highest level of similarity between two sequences and align these regions without regard for the alignment of rest of the regions. The two sequences to be aligned can be of different lengths.

Alignment Algorithms

DNA sequences are strings of letters from a four-letter alphabet called nucleotides (A, C, G, T). The length of a sequence is variable and sometimes we require the alignment of lengthy and highly variable or extremely numerous sequences. Hence, constructing algorithms to produce high-quality sequence alignments using four letters becomes a real challenge. In general, computational approaches to sequence alignment are classified as either global or local alignments. By global alignment, we consider aligning the entire scope of all query sequences against a reference sequence. On the other hand, the method of local alignment identifies isolated regions of high similarity within the entire sequence, which makes the technique a better choice in some situations but a more complex one in general.

Local Alignment: The Smith Waterman Algorithm

The Smith–Waterman algorithm is a well-known algorithm for performing local sequence alignment; that is, for determining similar regions between two nucleotide or protein sequences. Instead of looking at the total sequence, the Smith–Waterman algorithm compares segments of all possible lengths and optimizes the similarity measure. The explanation of the algorithm is given below:

The Smith-Waterman is used to compute and find the optimal local alignment or region which shares the same properties of two sequences. The procedure consists of two steps:

Step 1: Fill in the dynamic programming matrix.

Step 2: Find the maximal value (score) and trace back the patch that leads to maximal score to find the optimal local alignment.

The basis of a Smith-Waterman search is comparison of the two DNA sequences. It uses individual pair-wise comparison between characters such as:

$$For\ 0 \le i \le M, 0 \le j \le N$$

$$D_{io} = D_{oj} = 0$$

$$for\ 1 \le i \le M, and\ 1 \le j \le N,$$

$$D_{ij} = 0$$

or

$$D_{ij} = max \begin{cases} D_{(i-1)(j-1)} + Sbt_{(i,j)} \\ D_{(i-1)(j)} + d \\ D_{(i)(j-1)} + d \end{cases}$$

Where, d=penalty, Sbt =Substitution matrix, i = matrix cell row of search sequence, j= matrix cell column of search sequence, M=maximum length of target sequence, N= maximum length of target sequence, D(i,j)= dynamic matrix cell.

Global Alignment: The Needleman Wunsch Algorithm

Needleman-Wunsch uses dynamic programming in order to obtain global alignment between two sequences. Global alignment, as the name suggests takes into account all the elements of the two sequences while aligning the two sequences. We can also call it as an "end to end" alignment.

In Needleman-Wunsch algorithm, a scoring matrix of size m*n (m being the length of longer sequence and n being that of the shorter sequence) is first formed.

The optimal score at each matrix position is calculated by adding the current match score to previously scored positions and subtracting gap penalties. Each matrix position may have a positive, negative or 0 value.

For two sequences

S1=a1a2.................am (1.2)

S2=b1b2.................bn (1.3)

where

Tij=T(a1a2.................am, b1b2.................bn) then the element at the i,jth position of the matrix Tij is given by

$$Tij = Max \left\{ Ti - 1, j - 1 + s, \right.$$
$$Max(Ti - x, j - px), x >= 1$$
$$Max(Ti, j - y - py), y >= 1$$

Where, Tij is the score at position i in the sequence S1 and j in the sequence S2, T(ai,bj) is the score for aligning the characters at positions i and j, px is the penalty for a gap of length x in the sequence S1, and, py is the penalty for a gap of length y in the sequence S2.

Proposed DNA Alignment

For applying a global and local alignment and to get a score for both of them, the user can enter the sequence in two ways. The first way is by the accession numbers of the sequence to retrieve the sequences in its ORF (Open Reading Frames). The second way is to retrieve the sequences from the web (public database) and bringing the sequence information into the MATLAB environment. After that we can get global alignment (NW) and local alignment (SW) with a score that determines the degree of similarity. Dot plots are one of the easiest ways to find similarity between the two sequences. Many dots in the dot plot line up to form diagonal lines indicating good alignment between the two sequences. In the proposed work, the results are also presented in the form of dot plots.

Result Simulation and Discussion

Retrieving Sequences from a Database

Different sequences that have to be analyzed, aligned and read are retrieved from public database into MATLAB environment.

Open Reading Frame of Human DNA sequence

Figure shows the open reading frame of human. Once the ORF for a gene or mRNA is known, the user can translate a nucleotide sequence to its corresponding amino acid sequence.

Sequence Comparison by using Dot Plot

The most basic sequence alignment method is the dot matrix method also known as dot plot method. It is the graphical way of comparing two dimensional matrix. In dot matrix, two sequences to be compared are written in horizontal and vertical axes of the matrix. MATLAB function has been used for this comparison. When the two sequences have substantial regions of similarity, many dots line up to form diagonal lines, which reveals the sequence alignment.

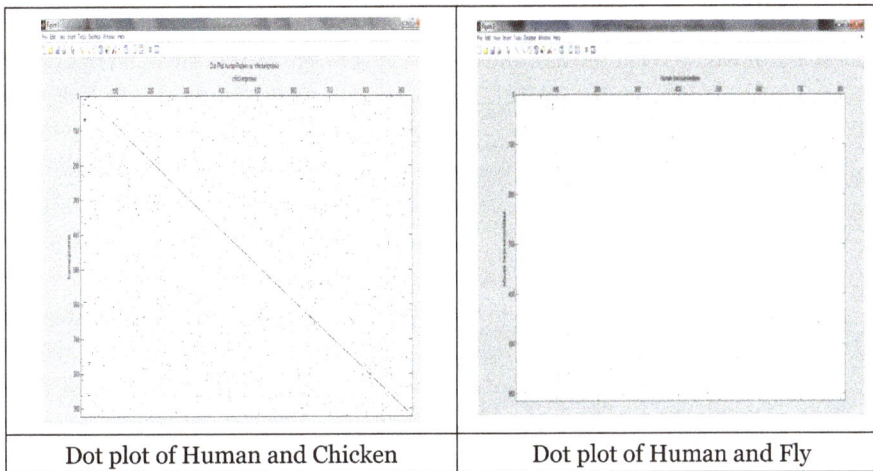

Dot plot of Human and Chicken	Dot plot of Human and Fly

Dot plots of human and chicken DNA sequences and human and fly DNA sequences have been shown in fig and fig respectively. The dot plots above shown shows that human and chicken DNA sequences show better alignment as compared to human and fly DNA sequences.

Global Alignment of Sequences by using Needleman Wunsch Algorithm

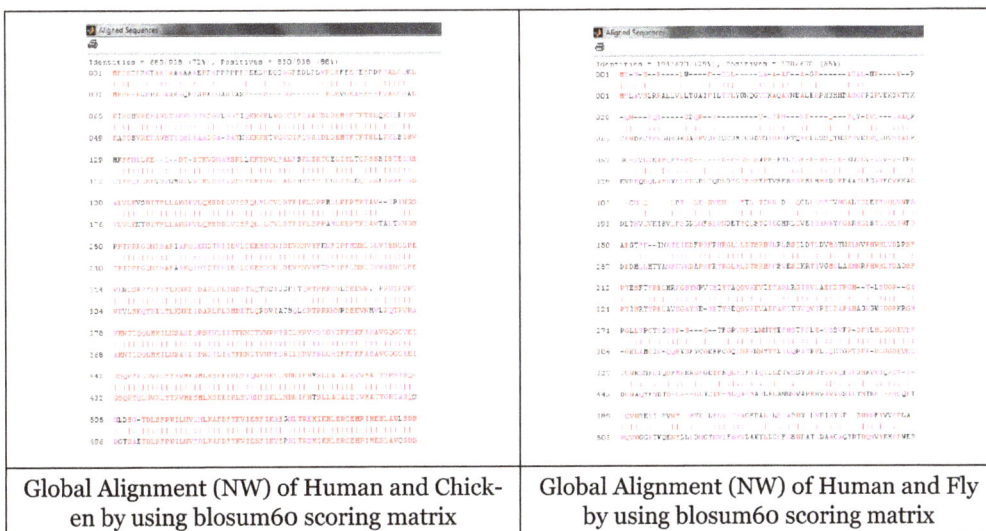

Global Alignment (NW) of Human and Chicken by using blosum60 scoring matrix	Global Alignment (NW) of Human and Fly by using blosum60 scoring matrix

Global alignment of human and Chicken DNA sequences is shown in fig and of human and fly DNA sequences is shown in fig respectively.

Local Alignment of Sequences by using Smith Waterman Algorithm

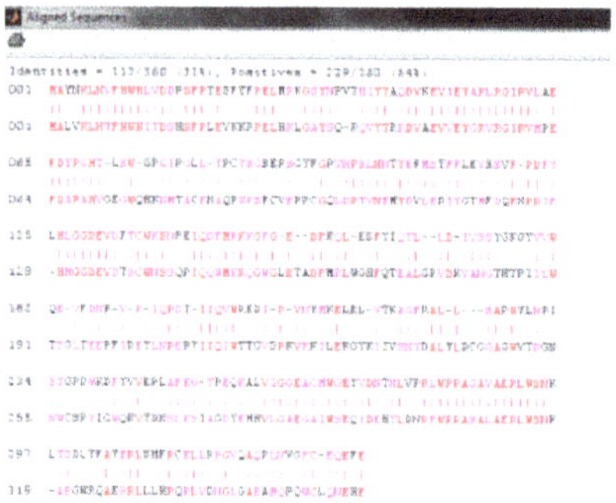

Local Alignment (SW) of Human and Fly by using blosum60 substitution matrices

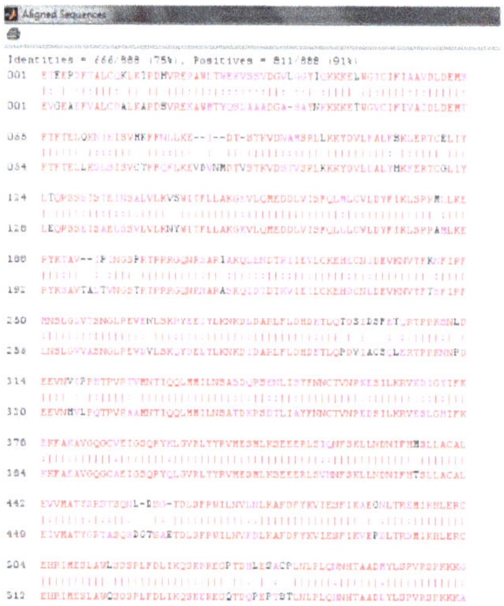

Local Alignment (SW) of Human and Chicken by using blosum60 substitution matrices

Local alignment of human and fly DNA sequences is shown in figure above and of human and chicken DNA sequences is shown in figure.

Probalign

Protein sequence alignment is likely the most commonly used task in bioinformatics. Applications include detecting functional regions in proteins and reconstructing complex evolutionary histories. Techniques for constructing accurate alignments are therefore of great interest to the bioinformatics community.

Methods

Posterior Probabilities and Maximal Expected Accuracy Alignment

Most alignment programs compute an optimal sum-of-pairs alignment or a maximum probability alignment using the Viterbi algorithm. An alternative approach is to search for the maximum expected accuracy alignment. The expected accuracy of an alignment is based upon the posterior probabilities of aligning residues in two sequences.

Consider sequences x and y and let $a*$ be their true alignment. Following the description in the posterior probability of residue xi aligned to yj in $a*$ is defined as

$$p(x_i \sim y_j \in a* \mid x,y) = \sum_{a \in A} P(a \mid x,y) 1\{x_i \sim y_j \in a\}$$

where A is the set of all alignments of x and y. $P(a|x,y)$ represents the probability (our belief) that alignment a is the true alignment $a*$. This can easily be calculated using a pairwise HMM if all the parameters are known. From hereon we represent the posterior probability as $P(x_i \sim y_j)$ with the understanding that it represents the probability of x_i aligned to y_j in the true alignment $a*$. Given the posterior probability matrix $P(x_i \sim y_j)$, we can compute the maximal expected accuracy alignment using the following recursion

$$A(i,j) = max \begin{cases} A(i-1,j-1) + P(x_i \sim y_j) \\ A(i-1,j) \\ A(i,j-1) \end{cases}$$

Probcons estimates posterior probabilities for amino acid residues using pair HMMs and unsupervised learning of model parameters. It then proceeds to construct a maximal expected accuracy alignment by aligning pairs of sequence profiles along a guide-tree followed by iterative refinement. In this investigation, we examine a different technique of estimating posterior probabilities; we use suboptimal alignments generated using the partition function of alignments.

According to equation above, as long as we have an ensemble of alignments A with their probabilities $P(a|,x,y)$ we can compute the posterior probability $P(x_i \sim y_j)$ by summing up the probabilities of alignments where x_i is paired with y_j. One way to generate an ensemble of such alignments is to use the partition function methodology.

Posterior Probabilities by Partition Function

Amino acid scoring matrices, normally used for sequence alignment, are represented as log-odds scoring matrices. The commonly used sum-of-pairs score of an alignment a is defined as the sum of residue-residue pairs and residue-gap pairs under an affine penalty scheme.

$$S(a) = T \sum_{(i,j) \in a} \text{In}(M_{ij} / f_i f_j) + (gap\_penalites)$$

Here T is a constant (depending upon the scoring matrix), M_{ij} is the mutation probability of residue i changing to j and f_i and f_j are background frequencies of residues i and j. In fact, it can be shown that any scoring matrix corresponds to a log odds matrix.

The probability of alignment $a, P(a)$, of sequences x and y can be defined as

$$P(a) \propto e^{S(a)/T}$$

where $S(a)$ is the score of the alignment under the given scoring matrix. In this setting one can then treat the alignment score as negative energy and T as the thermodynamic temperature, similar to what is done in statistical mechanics. Analogous to the statistical mechanical framework, the partition function of alignments is

$$Z(T) = \sum_{a \in A} e^{S(a)/T}$$

where A is the set of all alignments of x and y. With the partition function in hand, the probability of an alignment a can now be defined as

$$P(a,T) = e^{S(a)/T} / Z(T)$$

As T approaches infinity all alignments are equally probable, whereas at small values of T, only the nearly optimal alignments have the highest probabilities. Thus, the temperature parameter T can be interpreted as a measure of deviation from the optimal alignment. The alignment partition function can be computed using recursions similar to the Needleman-Wunsch dynamic algorithm. Let Z^M_{ij} represent the partition function of all alignments of $x_{1..i}$ and $y_{1..j}$ ending in xi paired with y_j, and $S_{ij}(a)$ represent the score of alignment a of $x_{1..i}$ and $y_{1..j-1}$. According to equation

$$Z^M_{i,j} = \sum_{a \in A_{ij}} e^{S_{ij}(a)/T} = \left(\sum_{a \in A_{i-1,j-1}} e^{S_{i-1,i-1}(a)/T} \right) e^{s(x_i, y_j)/T}$$

where A_{ij} is the set of all alignments of $x_{1..i}$ and $y_{1..j}$, and $s(x_i y_j)$ is the score of aligning residue x_i with y_j. The summation in the bracket on the right hand side of equation is precisely the partition function of all alignments of $x_{1..i-1}$ and $y_{1..j-1}$. We can thus compute the partition function matrices using standard dynamic programming.

$$Z^M_{i,j} = \left(Z^M_{i-1,j-1} + Z^E_{i-1,j-1} + Z^F_{i-1,j-1} \right) e^{s(x_i,y_j)/T}$$
$$Z^E_{i,j} = Z^M_{i,j-1} e^{g/T} + Z^E_{i,j-1} e^{ext/T}$$
$$Z^F_{i,j} = Z^M_{i-1,j} e^{g/T} + Z^F_{i-1,j} e^{ext/T}$$
$$Z_{i,j} = Z^M_{i,j} + Z^E_{i,j} + Z^F_{i,j}$$

Here $S(x,y)$ represents the score of aligning residue x_i with y_j, g is the gap open penalty, and ext is the gap extension penalty. The matrix Z^M_{ij} represents the partition function of all alignments ending in x_i paired with y_j. Similarly Z^E_{ij} represents the partition function of all alignments in which y_j is aligned to a gap and Z^F_{ij} all alignments in which xi is aligned to a gap. Once the partition function is constructed, the posterior probability of x_i aligned to y_j can be computed as

$$P(x_i \sim y_j) = \frac{Z^M_{i-1,j-1} Z'^M_{i+1,j+1}}{Z} e^{s(x_i,y_j)/T}$$

where $Z^{"M}_{i,j}$ is the partition function of alignments of subsequences $x_{i..m}$ and $y_{j..n}$ beginning with x_i paired with y_j and m and n are lengths of x and y respectively.

In equation $P(x_i \sim y_j) = \dfrac{Z^M_{i-1,j-1} Z'^M_{i+1,j+1}}{Z} e^{s(x_i,y_j)/T}$, $Z^M_{i-1,j-1}/Z$ and $Z'^M_{i+1,j+1}/Z$ represent the probabilities of all feasible suboptimal alignments (determined by the T parameter) of $x_{1..i-1}$ and $P(x,y)$, and $x_{i+1.m}$ and $y_{j+1.n}$ respectively, where m and n are lengths of x and y respectively. Thus, equation $P(x_i \sim y_j) = \dfrac{Z^M_{i-1,j-1} Z'^M_{i+1,j+1}}{Z} e^{s(x_i,y_j)/T}$ weighs alignments according to their partition function probabilities and estimates $P(x_i \sim y_j)$ as the sum of probabilities of all alignments where x_i is paired with y_j.

Recall the maximum expected accuracy alignment formulation described earlier. In order to compute such an alignment we need an estimate of the posterior probabilities. We utilize the partition function posterior probability estimates for constructing multiple alignments. For each sequence x,y in the input, we compute the posterior probability matrix $P(x_i \sim y_j)$ using equation. These probabilities are subsequently used to compute a maximal expected multiple sequence alignment using the Probcons methodology. First, the probabilistic consistency transformation is applied to improve the estimate of the probabilities. Briefly, the probabilistic consistency transformation is to reestimate the posterior probabilities based upon three-sequence alignments instead of pairwise. Note that this does not mean alignments are recomputed; our estimation (as done in Probcons) is still fundamentally based upon pairwise alignments. It is possible to compute a partition function of three-sequence alignments, and subsequently estimate posterior probabilities directly from them. However, in this proof of concept study, we examine performance on pairwise alignments only.

After the probabilistic consistency transformation, sequence profiles are next aligned in a post-order walk along a UPGMA guide-tree. As is commonly done, UPGMA guide trees are computed using pairwise expected accuracy alignment scores. Finally, iterative refinement is performed to improve the alignment.

We implement the Probalign approach by modifying the underlying Probcons program to read in arbitrary posterior probabilities for each pair of sequences in the input. All use of HMMs in the modified Probcons code is disabled. We modified the probA program of Muckstein et al., 2002 for computing partition function posterior probability estimates. The Probalign program, represented algorithmically, is a beta version and mainly for proof of concept.

Probalign Algorithm

1. For each pair of sequences (x,y) in the input set

    a. Compute partition function matrices $Z(T)$

    b. Estimate posterior probability matrix $P(x_i \sim y_j)$ for (x,y)

2. Perform the probabilistic consistency transformation and compute a maximal expected accuracy multiple alignment: align sequence profiles along a guide-tree and follow by iterative refinement.

Experimental Design

Alignment benchmarks: To test the accuracy of our method, we use three popular multiple protein sequence alignment benchmarks in the literature: BAliBASE, HOMSTRAD, and OXBENCH. BAliBASE is the most widely used benchmark for assessing protein multiple sequence alignments. Each alignment is well curated and contains core regions that represent reliable structurally alignable portions of the alignment. These alignable regions are used for evaluating accuracy and the remainder is ignored. BAliBASE 3.0 contains 5 sets of multiple protein alignments, each with different characteristics. RV11 contains 38 equidistant families with sequence identity less than 20%, while RV12 contains 44 equidistant families with sequence identity between 20% and 40%. Both of these lack sequences with large internal insertions (> 35 residues). RV20 contains 41 families with > 40% similarity and an orphan sequence which shares less than 20% similarity with the rest of the family. RV30 contains 30 families which contain sub-families with > 40% similarity but < 20% similarity across the sub-families. RV40 contains sequences with large N/C terminal extensions and is the largest set with 49 alignments, while RV50 contains sequences with large internal insertions and is the smallest with 16 alignments. Both RV40 and RV50 contain sequences that share > 20% similarity with at least one other sequence in the set. Overall, there are 217 benchmark alignments within BAliBASE 3.0.

HOMSTRAD is a curated database of structure-based alignments for homologous protein families. We use the April 2006 release for this study which contains 1033 families. HOMSTRAD contains all known protein structure clustered into homologous families and aligned on the basis of their 3-D structures.

OXBENCH is a set of structure-based alignments based on protein domains. It contains three sets of unaligned sequences: master, which are the unaligned protein domains in the true alignments; full, which contains full length unaligned proteins; and extended which contains additional proteins similar to the ones in unaligned master set. There are a total of 672 true master and extended alignments and 605 full sequence ones. Due to running time considerations, we exclude all datasets above 100 sequences.

Determining prediction accuracy: Given a true and estimated multiple sequence alignment, the accuracy of the estimated alignment is usually computed using two measures: the sum-of-pairs (SP) and the true column (TC) scores. SP is a measure of the number of correctly aligned residue pairs divided by the number of aligned residue pairs in the true alignment. TC is the number of correctly aligned columns divided by the number of columns in the true alignment. Both are standard measures of computing alignment accuracy.

Statistical significance: Statistically significant performance differences between the various alignment methods are calculated using the Friedman rank test, which is a standard measure used for discriminating alignments in benchmarking studies. Roughly speaking, the lower the reported P-value the less likely it is that the difference in ranking between the methods is due to chance. We consider P-values below 0.05 (a standard cutoff in statistics) to be statistically significant.

Programs compared and parameter settings. We compare Probalign to Probcons v1.1, MAFFT v5.851, and MUSCLE v3.6. We use the L-INS-i strategy of MAFFT, which is the most accurate according to latest benchmark tests by the MAFFT authors. The programs are compared using the scoring matrices and gap penalties recommended for their respective algorithms.

Probalign has two sets of parameters, one for the component that computes the posterior probabilities and the other for computing the maximal expected accuracy alignment. For the first component we use the Gonnet 160 scoring matrix with gap open and gap extension penalties set to -22 and -1 respectively. The default value of T (thermodynamic temperature) was set to 5 after comparing values 1 through 9 on BAliBASE RV11. For the second component, we use the exact same default parameters as that of Probcons, i.e. two rounds of probabilistic consistency and at most 100 rounds of iterative refinement.

Effect of Thermodynamic Temperature

We first look at the effect of different values of the thermodynamic temperature T on Probalign. Table below shows that T =5 is optimal on RV11. These settings of T appear to work well for the Gonnet 160 matrix and its affine gap penalties; therefore, we set T =5 for the remainder of our experiments.

T	Mean SP / TC	T	Mean SP / TC	T	Mean SP / TC
1	51.43 / 24.89	4	65.23 / 43.03	7	60.28 / 36.58
2	55.06 / 29.08	5	69.32 / 45.26	8	49.51 / 25.76
3	57.90 / 32.39	6	66.18 / 40.87	9	41.12 / 18.84

Effect of different thermodynamic temperatures on Probalign on RV11 subset of BAliBASE 3.0.

Benchmark Comparisons

In table below we compare mean SP scores and TC of Probalign to other methods on BAliBASE 3.0. Probalign averages are the high- est on the RV11, RV12, and RV40 subsets, as well as the full BAliBASE dataset. MAFFT does better on the remaining three datasets. Although the differences are small, Probalign ranks statis- tically significantly higher than all three methods on RV12, RV40, and the full BAliBASE dataset. No method ranked statistically significantly higher than Probalign on any of the BAliBASE subsets.

Data	Probalign	MAFFT	Probcons	MUSCLE RV11
RV12	94.6 / 86.2	93.6 / 83.8	94.1 / 85.5	91.7 / 80.4
RV20	92.6 / 43.9	92.7 / 45.3	91.7 / 40.6	89.2 / 35.1
RV30	85.2 / 56.4	85.6 / 56.9	84.5 / 54.4	80.3 / 38.3
RV40	92.2 / 60.3	92.0 / 59.7	90.3 / 53.2	86.7 / 47.1
RV50	89.3 / 55.2	90.0 / 56.2	89.4 / 57.3	85.7 / 48.7
All	87.6 / 58.9	87.1 / 58.6	86.4 / 55.8	82.5 / 48.5

Mean SP / TC scores on BAliBASE 3.0.

Method	RV11	RV12	RV20	RV30	RV40	RV50	All
MAFFT	NS	< 0.005	NS	NS	< 0.005	NS	< 0.005
Probcons	0.049	0.0233	NS	NS	< 0.005	NS	< 0.005
MUSCLE	< 0.005	< 0.005	0.008	< 0.005	< 0.005	NS	< 0.005

P-values of Friedman rank test on BAliBASE TC scores. In all cases of statistical significance (< 0.05) Probalign is ranked higher. NS indicates non statistically significant.

We also test Probcons by retraining (on BAliBASE 3.0) with single and pair emission probabilities set to the background and mutation matrix probabilities of Gonnet 160. In this way we can test if the Probalign improvements are purely a result of scoring matrix differences. The performance of

Probcons performance does not improve. In fact, it actually does worse than with training on the (default) Blosum 62 matrix.

Table below compares the CPU running time of Probalign to the other methods on RV11 and RV12 subsets of BAliBASE. While Probalign is the slowest, its running time is still tractable. Our current beta implementation is a pipeline of C++ programs and Perl scripts linked by system calls. An integrated version (which is in progress) will yield a much faster implementation.

Data	Probalign	MAFFT	Probcons	MUSCLE
RV11	6.64	0.98	3.65	0.71
RV12	17.73	1.28	10.46	0.74

Mean CPU time (in seconds) on RV11 and RV12 subsets of BAliBASE 3.0.

Finally, table below compares mean SP and TC scores on the HOMSTRAD and OXBENCH benchmarks. Probalign mean SP and TC scores rank highest on HOMSTRAD, OXBENCH, and OX-BENCH-full with P-value < 0.005. Moreover, on the OXBENCH-extended dataset, no method ranked statistically significantly higher than Probalign. In fact, Probalign ranks higher than Probcons on OXBENCH-extended with P-value 0.014

Data	Probalign	MAFFT	Probcons	MUSCLE
HOMSTRAD	82.2 / 77.9	80.4 / 75.9	81.9 / 77.4	80.8 / 76.3
OXBENCH	89.8 / 85.1	88.4 / 83.2	89.3 / 84.2	89.4 / 84.4
OXBENCH (full)	84.0 / 77.0	82.8 / 75.3	83.2 / 75.7	82.6 / 74.8
OXBENCH (extend)	92.0 / 89.6	92.5 / 90.0	92.4 / 89.8	91.8 / 89.0

Mean SP / TC scores on HOMSTRAD and OXBENCH

Simulation of N/C Terminal Extensions

Probalign's performance improvement is most significant over all methods on the RV40 subset of BAliBASE. Recall that this dataset contains sequences with long N/C terminal extensions. We rely on simulation to further test Probalign's improvement on this type of data. We begin by computing the maximum parsimony model trees (with edge lengths) on arbitrary selected alignments from the RV11 subset of BAliBASE 3.0. We select the BB11003, BB11004, BB11008, BB11009, and BB11010 alignments, all of which contain four sequences and branch length ranging from conservative to divergent. For each tree, we generate a root protein sequence with the same background probability distribution as Dayhoff's. We define core regions of this sequence as randomly selected contiguous region (with probability 0.25) ranging from length 1 to 30 (with uniform probability). We then evolve sequences using the ROSE model. However, in the defined core regions, the mutation probability is reduced (by half) and no insertion deletions are allowed.

Briefly, ROSE interprets each branch length as PAM units of evolution. On a branch of length k, the probability of substitution is given by Mk where M is the PAM1 mutation probabilities. For insertion (or deletion) it randomly picks an amino acid with prob- ability *insert_threshold * branch_length * sequence_length* and inserts (or deletes) a sequence of length given by an exponential distribution. Once the simulated sequences are generated, we attach a randomly generated sequence to each end of each sequence with probability 0.25, which constitute our artificial N/C extensions.

For each model tree, we produce a root sequence of length 100, and the (insertion, deletion) thresholds are set to (0.0005, 0.000125), meaning the deletion threshold is $1/4^{th}$ the insertion. We generate 100 sequence sets for each model tree, and align using Probalign, MAFFT, and Probcons. The alignments are compared against the core regions of the true alignment (known by simulation). Table below shows that Probalign wins for all model trees.

Probalign SP and TC scores also rank higher than all methods with P-value < 0.05 (except for BB11009 where all methods do equally well). We also examined performance on simulated data containing long internal insertions, along with the N/C extensions, and saw similar results (data not shown).

Model tree	Probalign	MAFFT	Probcons
BB11003 (164)	77.1 / 63.7	72.7 / 58.2	72.4 / 56.9
BB11004 (132)	89.5 / 83.0	86.7 / 78.3	86.8 / 78.5
BB11008 (92)	97.9 / 95.9	96.8 / 93.9	96.5 / 93.3
BB11009 (33)	99.8 / 99.7	99.8 / 99.7	99.8 / 99.6
BB11010 (184)	63.4 / 46.9	58.1 / 41.0	60.1 / 414

Mean SP / TC scores on different model trees. Also shown are average branch lengths (PAM units of evolution) for each model tree.

Datasets with long and variable length sequences

Not only does the RV40 subset contain sequences with large N/C extension, but they are also highly variable in length. In fact, many constituent proteins are at least 1000 residues in length. Based on our results thus far, we conjecture that Probalign does best when presented such datasets. To test this hypothesis, we select all unaligned datasets in BAliBASE 3.0 where the standard deviation in sequence length is at least 100 or 200 and the maximum length is at least 500 or 1000. For these four possible permutations, we compare the mean SP and TC scores of Probalign to the other methods.

Max length / Standard dev.	Probalign	MAFFT	Probcons	MUSCLE
500 / 100	88.4 / 56.6	88.0 / 58.0	86.7 / 51.6	81.5 / 42.5
500 / 200	88.5 / 54.6	87.0 / 51.9	87.2 / 48.9	81.9 / 42.4
1000 / 100	91.4 / 58.1	90.4 / 55.7	89.7 / 51.6	84.3 / 44.1
1000 / 200	90.7 / 55.0	89.3 / 51.4	89.2 / 48.7	83.2 / 42.5

Mean SP / TC scores on BAliBASE 3.0 datasets with standard deviation of length at least 100 and 200 and maximum sequence length at least 500 and 1000.

Table shows that the improvement of Probalign over other methods increases as both the standard deviation of the mean length and the maximum sequence length increases. The Probalign mean column score (TC) is 2.8%, 2.4%, and 3.7% better than MAFFT at the 500/200, 1000/100, and 1000/200 settings, respec- tively, and at least 5% better than Probcons on all four combina- tions. Furthermore, even though the mean TC is lower than that of MAFFT in row one, *Probalign ranked higher than all methods on each of the four settings with P-value < 0.005 (for both TC and SP scores).*

Table shows mean SP and TC scores broken down for each BAliBASE subset but containing only those datasets with maximum sequence length at least 1000 and standard deviation of length at least 100 and 200. We omit MUSCLE from this comparison since it is poorest on these types of datasets. At the 1000/100 setting, Probalign mean TC score is at least 2.8%, 3%, and 4% better than MAFFT and Probcons on RV12, RV30, and RV40 subsets, respectively. At the 1000/200 setting, TC improvement on both RV30 and RV40 increases to at least 5%. However, only on RV40 is Probalign statistically significantly ranked highest for both SP and TC score (with P-value < 0.005). No method ranked statistically significantly higher than Probalign.

Table Mean SP / TC scores for datasets with max sequence length at least 1000 and standard deviation of length at least 100 and 200 for each BAliBASE subset. The number of datasets in each BAliBASE subset (RV11 through RV50) satisfying these criteria is indicated in parentheses.

Max length / Standard dev.	Probalign	MAFFT	Probcons
RV11 1000 / 100 (1)	62.5 / 39.0	55.2 / 36.0	62.8 / 38.0
1000 / 200 (1)	62.5 / 39.0	55.2 / 36.0	62.8 / 38.0
RV12 1000 / 100 (5)	93.6 / 81.6	91.5 / 77.0	92.3 / 78.8
1000 / 200 (5)	93.6 / 81.6	91.5 / 77.0	92.3 / 78.8
RV20 1000 / 100 (6)	92.3 / 42.0	91.7 / 41.0	91.0 / 38.5
1000 / 200 (5)	91.6 / 34.6	90.9 / 34.0	90.1 / 30.4
RV30 1000 / 100 (3)	90.8 / 67.3	90.6 / 64.3	89.4 / 63.3
1000 / 200 (1)	77.2 / 40.0	76.1 / 34.0	73.6 / 35.0
RV40 1000 / 100 (25)	92.7 / 59.3	91.0 / 54.8	89.9 / 48.2
1000 / 200 (20)	93.0 / 57.3	90.8 / 52.1	90.6 / 47.6
RV50 1000 / 100 (6)	88.1 / 48.5	91.2 / 55.8	89.7 / 52.2

On RV50, MAFFT is the winner on both the full dataset and on the subsets in Table 8, but not statistically signifi- cantly ranked higher. By reducing the gap extension penalty (to allow for large internal insertions), Probalign's TC score improves considerably (but not statistically significantly) as shown in Table below. The TC score with 0.2 gap extension penalty is 3.2% better than Probcons and MAFFT at the 1000/200 setting.

Table 9. Mean SP / TC scores for the full RV50 BAliBASE dataset (long internal insertions) in row two and for RV50 datasets with long and hetero- geneous length sequences (last two rows). The number of datasets meeting these criteria is indicated in parentheses.

RV50 Dataset	Probalign	Probalign	MAFFT	Probcons
	(gap ext 0.2)	(gap ext 1.0)		
Complete	87.8 / 56.4	89.3 / 55.2	**90.0** / 56.2	89.4 / **57.3**
Max len / Std dev				
1000/100 (6)	88.2 / **56.0**	88.1 / 48.5	**91.2** / 55.8	89.7 / 52.2
1000/200 (4)	85.9 / **49.0**	85.0 / 43.5	**89.1** / 45.8	87.3 / 45.8

We perform one more test here to examine performance on heterogeneous length sequences. We consider reference set 6 of BAliBASE 2.0 containing repeats. Repeats are much smaller than the original sequence and most of the repeat datasets containing highly variable length sequences. Reference 6 of BAliBASE contains 13 reference alignments of repeats and several more repeat datasets classified into six different subsets. We gather all datasets in reference 6 (for a total of 77) and considered only those with maximum sequence length at least 500 and 1000, and standard deviation of length at least 100, 200, 300, and 400. Again, we omit MUSCLE because it performs worse than the three other methods on this type of data.

Table: Mean SP / TC scores on BAliBASE 2.0 reference 6 (repeat) datasets with std. deviation of length at least 100, 200, 300, and 400, and maximum sequence length at least 500 and 1000. Indicated in parentheses are the number of datasets meeting these conditions.

Max length / Standard dev. 500 / 100 (40)	Probalign	MAFFT	Probcons
	89.1 / 44.9	87.3 / 49.0	87.4 / 38.6
500 / 200 (21)	88.3 / 43.8	85.0 / 46.4	86.7 / 40.0
500 / 300 (9)	95.3 / 61.0	82.6 / 51.3	87.3 / 46.6
500 / 400 (5)	94.6 / 55.0	72.0 / 38.2	79.8 / 38.0
1000 / 100 (15)	90.2 / 43.3	82.4 / 36.9	85.4 / 27.6
1000 / 200 (12)	89.2 / 38.2	79.7 / 32.4	83.6 / 27.7
1000 / 300 (7)	94.5 / 52.8	78.3 / 42.4	83.9 / 34.6
1000 / 400 (5)	94.6 / 55.0	72.0 / 38.2	79.8 / 38.0

The Probalign improvements on these datasets are the largest observed so far. As the maximum sequence length and the standard deviation in length increases so does the Probalign improvement. When standard deviation of length is at least 300 and 400, Probalign SP and TC score is at least 10% and 15% better than the next best method. While no method is ranked statistically significantly better than any other on these datasets, these large Probalign improvements gained warrant significant merit.

Alignment-free Sequence Analysis

Alignment-free sequence analyses have been applied to problems ranging from whole-genome phylogeny to the classification of protein families, identification of horizontally transferred genes, and detection of recombined sequences. The strength of these methods makes them particularly useful for next-generation sequencing data processing and analysis. However, many researchers are unclear about how these methods work, how they compare to alignment-based methods, and what their potential is for use for their research.

Alignment-free approaches to sequence comparison can be defined as any method of quantifying sequence similarity/dissimilarity that does not use or produce alignment (assignment of residue–residue correspondence) at any step of algorithm application. From the start, such restriction places the alignment-free approaches in a favorable position—as alignment-free methods do not rely on dynamic programming, they are computationally less expensive (as they are generally of a linear complexity depending only on the length of the query sequence) and therefore suitable for whole genome comparisons. Alignment-free methods are also resistant to shuffling and recombination events and are applicable when low sequence conservation cannot be handled reliably by alignment. Finally, in contrast to alignment-based methods, they do not depend on assumptions regarding the evolutionary trajectories of sequence changes. Although these characteristics apply to all alignment-free methods, there are more than 100 techniques to consider.

Alignment-free approaches can be broadly divided into two groups : methods based on the frequencies of subsequences of a defined length (word-based methods) and methods that evaluate the informational content between full-length sequences (information-theory based methods). There are also methods that cannot be classified in either of the groups, including those based on the length of matching words (common, longest common, or the minimal absent words between sequences), chaos game representation, iterated maps, as well as graphical representation of DNA sequences, which capture the essence of the base composition and distribution of the sequences in a quantitative manner.

All of the alignment-free approaches are mathematically well founded in the fields of linear algebra, information theory, and statistical mechanics, and calculate pairwise measures of dissimilarity or distance between sequences. Conveniently, most of these measures can be directly used as an input into standard tree-building software, such as Phylip or MEGA.

Working of Word Frequency-based Methods

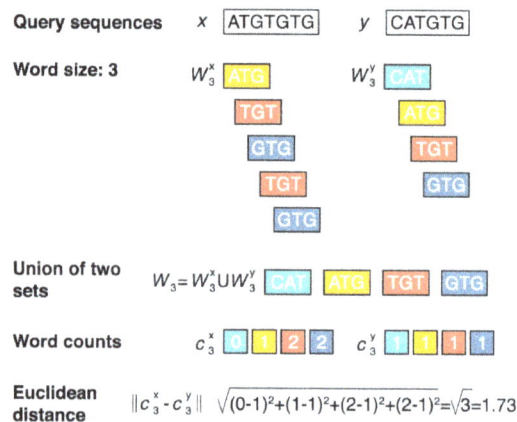

Alignment-free calculation of the word-based distance between two sample DNA sequences ATGTGTG and CATGTG using the Euclidean distance

The rationale behind these methods is simple: similar sequences share similar words/k-mers (subsequences of length k), and mathematical operations with the words' occurrences give a good relative measure of sequence dissimilarity. The method is also tightly coupled with the idea of genomic signatures, which were first introduced for dinucleotide composition (e.g., GC content) and further extended to longer words. This process can be broken into three key steps.

First, the sequences being compared must be sliced up into collections of unique words of a given length. For example, two DNA sequences $x =$ ATGTGTG and $y =$ CATGTG and a word size of three nucleotides (3-mers) produces two collections of unique words:

$W_3^X = \{ATG,\ TGT,\ GTG\}$ and $W_3^Y = \{CAT,\ ATG,\ TGT,\ GTG\}$. Because some words are often present in one sequence but not in the other sequence (i.e., CAT in y but not in x), we create a full set of words that belong to at least W_3^X or W_3^Y to further simplify the calculations, resulting in the union set $W_3 = \{ATG,\ CAT,\ GTG,\ TGT\}$.

The second step is to transform each sequence into an array of numbers (vector) (e.g., by counting the number of times each particular word (from W_3) appears within the sequences). For sequences x and y, we identify two real-valued vectors: $c_3^X = (1,\ 0,\ 2,\ 2)$ and $c_3^Y = (1,\ 1,\ 1,\ 1)$.

The last step includes quantification of the dissimilarity between sequences through the application of a distance function to the sequence-representing vectors c_3^X and c_3^Y. This difference is very commonly computed by the Euclidean distance, although any metric can be applied. The higher the dissimilarity value, the more distant the sequences; thus, two identical sequences will result in a distance of 0.

Word-based alignment-free algorithms come in different colors and flavors, with methodological variations at each of the three basic steps. In the first step, one can try any resolutions of word lengths—it is important to choose words that are not likely to commonly appear in a sequence (the shorter the word, then the more likely it will appear randomly in a sequence). In practice, the word size (k) of 2–6 residues produces stable and optimal protein sequence comparisons across a wide range of different phylogenetic distances; in nucleotide sequence analyses, k can safely be set to 8–10 for genes or RNA, 9–14 bases for general phylogenetic analyses, and up to 25 bases in case of comparison of isolates of the same bacterial species. As a rule of thumb, smaller k-mers should be used when sequences are obviously different (e.g., they are not related) whereas longer k-mers can be used for very similar sequences. Alternatively, DNA/RNA or protein alphabet can be reduced to a smaller number of symbols based on chemical equivalences. This procedure may increase the detection of homologous sequences that display very low identity. For example, the four-letter DNA alphabet can be distilled to two-letter purine–pyrimidine encoding, and proteins can be represented by 5, 4, 3, or even 2 letters according to their different physical–chemical properties. The second step (mapping sequences onto vectors) is by far the most customizable; instead of using vectors of word counts or word frequencies, there are many other ways to create vectors, which range from weighting techniques to normalization and clustering. Additionally, because word-based methods operate on vectors, their mathematical elegance allows the employment of more than 40 functions other than the Euclidean distance, such as the Pearson correlation coefficient, Manhattan distance, and Google distance.

Working of Information Theory-based Methods

Information theory-based methods recognize and compute the amount of information shared between two analyzed biological sequences. Nucleotide and amino acid sequences are ultimately strings of symbols, and their digital organization is naturally interpretable with information theory tools, such as complexity and entropy.

For example, the Kolmogorov complexity of a sequence can be measured by the length of its shortest description. Accordingly, the sequence AAAAAAAAAA can be described in a few words (10 repetitions of A), whereas CGTGATGT presumably has no simpler description than specification nucleotide by nucleotide (1 C, then 1 G and so on). Intuitively, longer sequence descriptions indicate more complexity. However, Kolmogorov did not address the method to find the shortest description of a given string of characters. Therefore, the complexity is most commonly approximated by general compression algorithms (e.g., as implemented in zip or gzip programs) where the length of a compressed sequence gives an estimate of its complexity (i.e., a more complex string will be less compressible). The calculation of a distance between sequences using complexity (compression) is relatively straightforward . This procedure takes the sequences being compared (x = ATGTGTG and y = CATGTG) and concatenates them to create one longer sequence (xy = ATGTGTGCATGTG). If x and y are exactly the same, then the complexity (compressed length) of xy will be very close to the complexity of the individual x or y. However, if x and y are dissimilar, then the complexity of xy(length of compressed xy) will tend to the cumulative complexities of x and y. Of course, there are as many different information-based distances as there are methods to calculate complexity. For example, Lempel–Ziv complexity is a popular measure that calculates the number of different subsequences encountered when viewing the sequence from beginning to end . Once the complexities of the sequences are calculated, a measure of their differences (e.g., the normalized compression distance) can be easily computed. Many DNA-specific compression algorithms are currently being applied to new types of problems.

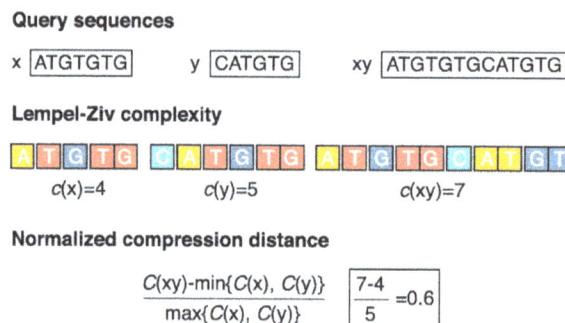

Query sequences

x ATGTGTG   y CATGTG   xy ATGTGTGCATGTG

Lempel-Ziv complexity

A T G TG   C A T G TG   A T G TG C A T G T

$c(x)=4$ $c(y)=5$ $c(xy)=7$

Normalized compression distance

$$\frac{C(xy)-\min\{C(x),\,C(y)\}}{\max\{C(x),\,C(y)\}} \qquad \frac{7-4}{5}=0.6$$

Above figure shows alignment-free calculation of the normalized compression distance using the Lempel–Ziv complexity estimation algorithm. Lempel–Ziv complexity counts the number of different words in sequence when scanned from left to right (e.g., for s = ATGTGTG, Lempel–Ziv complexity is 4: A|T|G|TG). Description of compression algorithms in alignment-free analysis has been reviewed extensively

Another example of an information measurement often applied to biological sequences is entropy. This measurement is not similar to the entropy referenced in thermodynamics. The concept of Shannon entropy came from the observation that some English words, such as "the" or "a", are very frequent and thus unsurprising. Thus, these words are redundant because the message can probably be understood without them. The real essence of the message comes from words that are rare, such as "treasure" or "elixir". Therefore, Shannon developed a formula to quantify the uncertainty (entropy) of finding a given element (word) in an analyzed sequence (text). Using Shannon's concept, Kullback and Leibler introduced a relative entropy measure (Kullback–Leibler divergence, KL) that allowed for a comparison of two sequences. The procedure involves the calculation of the frequencies of symbols or words in a sequence and the summation of their entropies in the compared sequences.

Both information-theory concepts (complexity and entropy) have a clear association despite their methodical differences. For instance, a low-complexity sequence (e.g., AAAAAAAAA) will have smaller entropy than a more complex sequence (e.g., ACCTGATGT). The application of information theory in the field of sequence analysis and comparison has exploded in recent years, ranging from global (block entropies and coverage) to local genome analyses (transcription factor binding sites, sequences as time-series and entropic profiles). Additionally, retrieving higher-level correlations in gene mapping and protein–protein interaction networks and the striking resemblance with communication systems is attracting research attention to this field.

Biological Sequence Format

A biological sequence is a single, continuous molecule of nucleic acid or protein. It can be thought of as a multiple inheritance class hierarchy. One hierarchy is that of the underlying molecule type: DNA, RNA, or protein. The other hierarchy is the way the underlying biological sequence is represented by the data structure. It could be a physical or genetic map, an actual sequence of amino acids or nucleic acids, or some more complicated data structure building a composite view from other entries.

FASTA Format

FASTA format is a text-based format for representing either nucleotide sequences or peptide sequences, in which base pairs or amino acids are represented using single-letter codes. A sequence in FASTA format begins with a single-line description, followed by lines of sequence data. The description line is distinguished from the sequence data by a greater-than (">") symbol in the first column. It is recommended that all lines of text be shorter than 80 characters in length.

An example sequence in FASTA format is:

```
>gi|186681228|ref|YP_001864424.1|  phycoerythrobilin:ferredoxin  oxidore-
ductase

MNSERSDVTLYQPFLDYAIAYMRSRLDLEPYPIPTGFESNSAVVGKGKNQEEVVTTSYAFQTAKLRQIR
A

AHVQGGNSLQVLNFVIFPHLNYDLPFFGADLVTLPGGHLIALDMQPLFRDDSAYQAKYTEPILPIFHAH
Q

QHLSWGGDFPEEAQPFFSPAFLWTRPQETAVVETQVFAAFKDYLKAYLDFVEQAEAVTDSQNLVAIKQA
Q

LRYLRYRAEKDPARGMFKRFYGAEWTEEYIHGFLFDLERKLTVVK
```

Sequences are expected to be represented in the standard IUB/IUPAC amino acid and nucleic acid codes, with these exceptions:

- lower-case letters are accepted and are mapped into upper-case;

- a single hyphen or dash can be used to represent a gap of indeterminate length;

- in amino acid sequences, U and * are acceptable letters (see below).

- any numerical digits in the query sequence should either be removed or replaced by appropriate letter codes (e.g., N for unknown nucleic acid residue or X for unknown amino acid residue).

The nucleic acid codes are:

```
A --> adenosine          M --> A C (amino)

C --> cytidine           S --> G C (strong)

G --> guanine            W --> A T (weak)

T --> thymidine          B --> G T C

U --> uridine            D --> G A T

R --> G A (purine)       H --> A C T

Y --> T C (pyrimidine)   V --> G C A

K --> G T (keto)         N --> A G C T (any)

                         - gap of indeterminate length
```

The accepted amino acid codes are:

```
A ALA alanine                        P PRO proline

B ASX aspartate or asparagine        Q GLN glutamine

C CYS cystine                        R ARG arginine

D ASP aspartate                      S SER serine

E GLU glutamate                      T THR threonine

F PHE phenylalanine                  U   selenocysteine

G GLY glycine                        V VAL valine

H HIS histidine                      W TRP tryptophan

I ILE isoleucine                     Y TYR tyrosine

K LYS lysine                         Z GLX glutamate or glutamine

L LEU leucine                        X   any

M MET methionine                     *   translation stop

N ASN asparagine                     -   gap of indeterminate length
```

FASTQ Format

Most modern sequencers produce FASTQ files as output, which is a modified version of a traditional FASTA formatted file. FASTQ files are ASCII text files that encode both nucleotide calls as well as 'quality information', which provides information about the confidence of each nucleotide. FASTQ format uses 4 lines for each read produced by the sequencer. Fastq files are nomally given the file extension ".fq" or ".fastq". A typical files looks something like this:

@SRR566546.970 HWUSI-EAS1673_11067_FC7070M:4:1:2299:1109 length=50

TTGCCTGCCTATCATTTTAGTGCCTGTGAGGTGGAGATGTGAGGATCAGT

+SRR566546.970 HWUSI-EAS1673_11067_FC7070M:4:1:2299:1109 length=50

hhhhhhhhhhghhghhhhhfhhhhhhfffffe`ee[`X]b[d[ed`[Y[^Y

@SRR566546.971 HWUSI-EAS1673_11067_FC7070M:4:1:2374:1108 length=50

GATTTGTATGAAAGTATACAACTAAAACTGCAGGTGGATCAGAGTAAGTC

+SRR566546.971 HWUSI-EAS1673_11067_FC7070M:4:1:2374:1108 length=50

hhhhgfhhcghghggfcffdhfehhhhcehdchhdhahehffffde`bVd

@SRR566546.972 HWUSI-EAS1673_11067_FC7070M:4:1:2438:1109 length=50

TGCATGATCTTCAGTGCCAGGACCTTATCAAGCGGTTTGGTCCCTTTGTT

+SRR566546.972 HWUSI-EAS1673_11067_FC7070M:4:1:2438:1109 length=50

Dhhhgchhhghhhfhhhhhdhhhhehhghfhhhchfddffcffafhfghe

The example above encodes 3 reads (each uses 4 lines to report information). Each read has:

1. Header line - This line must start with "@", followed by the name of the read (should be unique)

2. The nucleotide sequence

3. A second header line - This line must start with "+". Usually, the information is the same as in the first header line, but it can also be blank (The "+" is still required though)

4. Quality Information - For each nucleotide in the sequence, an ASCII encoded quality score is reported. The idea is that better quality scores indicate the base is reliably reported, while poor quality scores reflect uncertaintly about the true identity of the base.

These files represent the primary data generated by the sequencer, and will be requested by other researchers after you publish your study. Do not loose or modify these files If you are using Life Tech/ABI Solid sequencers, the data may be returned as a color space FASTQ file (usually with a *.csfastq" extension).

Duplicate and Archive

The most important thing you can do once you get primary data off the sequencer is back to it up. If you loose a mapping file (sam/bam), it's not a big deal as you can always remap your data, but if you lose the raw sequencing file (i.e. FASTQ), you're in trouble, and may have to repeat the experiment. Therefore, it is critical that you backup your files to a secure server that is [ideally] in a separate physical location than your primary copy. Storing a single copy on a RAID device does NOT count.

The other under appreciated task is to annotate your files. This means giving them proper names, and including information about how the experiment was done, cell type, etc. so that when you find the file later, you know what the experiment was. We would strongly recommend including a date in the file name, the cell type, the treatment, the type of experiment, the lab/scientist who performed the experiment. For example, consider naming them like this:

Lab-YYMMDD-Cell-Txn-expID.fastq

ChuckNorrisLab-120927-Bcell-LPS-0123.fastq

Also, it's probably a good idea to zip the files such that they take up less space. For example:

gzip *.fq *.Fastq

Removing Barcodes

Depending on your sequencing strategy, you may need to remove certain parts of the sequence that is not biologically meaningful. For example, if you sequence short RNAs that are between 15-40 bp in size, and you sequence them using 50 nucleotide reads, the sequence will start to identify the adapter used in sequencing library construction at the end of the RNA sequence.

Quality Value Encoding

Different version of the Illumina pipeline (from back in the day) can produce different encoding of quality. In general, recent implementations of the Illumina pipeline output Sanger-style quality encoding, so you should have to worry much about it. Many programs, such as bowtie for read mapping, have options to specify which style of encoding is used.

Performing Quality Controls on FASTQ files

It's a good idea to perform a general quality control check on your sequence files - this can help indicate if there were any major technical issues with your sequencing. A nice tool for this is FASTQC.

Usually, the easiest way to run FASTQC is on the command line:

mkdir OutputDirectory/

fastqc -o OutputDirectory/ inputFile.fastq

Unfortunately, it will complain if you do not create the output directory ahead of time. This analysis will produce several interesting analyses that help you understand how your sequencing went:

Manipulating FASTQ Files

Since most of the applications covered here involve "re-sequencing" of a known genome, the quality information about each base is not terribly important (it's more important when trying to identify SNP or for de novo genome assembly). However, sometimes, as a read is sequenced, errors start to appear and the reliability of the sequence goes down. In these cases it's best to remove these sequences from the mapping to improve downstream analysis.

An excellent resource for the manipulation of FASTQ files is the FASTX program suite. These programs can be very useful. In addition, some mapping tools (i.e. bowtie) have options that perform on-the-fly FASTQ manipulation, such as trimming from the 3' end.

Trimming Adapter Sequences from your Fastq Files

If you perform short RNA sequencing or another type of experiment where the functional sequences you are measuring might be smaller than the read length, it is likely that the 3' end of the read will be adapter sequences from the Illumina library preparation, and not relevant biological/genomic sequence. You must remove this sequence before trying to map or assemble the reads. The FASTX program fastx_clipper can perform adapter clipping.

Trimming Sequences Based on Quality Scores

If your reads are very long, you may want to trim sequences where the quality scores took a dive. This may be necessary for 100 bp reads if the last 20 bp are all random base calls. In this case the read may be hard to map since the final 20 bp will be largely wrong. The FASTX tool fastq_quality_trimmer is useful for this purpose.

SAM

The Sequence Alignment/Map (SAM) format is a generic nucleotide alignment format that describes the alignment of query sequences or sequencing reads to a reference sequence or assembly. Importantly:

- It is flexible enough to store all the alignment information generated by various alignment programs;

- It is simple enough to be easily generated by alignment programs or converted from existing alignment formats;

- It is compact in file size;

- It allows most of the operations on the alignment to work on a stream without loading the whole alignment into memory;

- It allows the file to be indexed by genomic position to efficiently retrieve all reads aligning to a locus.

SAM is a tab-delimited text format. SAM is a bit slow to parse; so there is a binary equivalent to SAM, called BAM.

SAM allows optional fields to be stored. In SAM, each alignment must contain a fixed number of mandatory fields that describe the key information about the alignment (such as coordinate detailed alignment and sequences) and may contain a variable number of optional fields which are less important or aligner specific.

Using SAM to Store Various Types of Alignments

SAM is able to store clipped alignments, spliced alignments, multi-part alignments, padded alignments and alignments in color space. The extended CIGAR string is the key to describing these types of alignments.

Clipped alignment

In Smith-Waterman alignment, a sequence may not be aligned from the first residue to the last one. Subsequences at the ends may be clipped off. We introduce operation 'S' to describe clipped alignment. Suppose the clipped alignment is:

clipped_alignment

```
REF:  AGCTAGCATCGTGTCGCCCGTCTAGCATACGCATGATCGACTGTCAGCTAGTCAGACTAGTCGATCGATGTG

READ:      gggGTGTAACC-GACTAGgggg
```

where on the read sequence, bases in uppercase are matches and bases in lowercase are clipped off. The CIGAR for this alignment is: 3S8M1D6M4S (which I interpret as 3 soft, 8 match, 1 deletion, 6 match and 4 soft).

Spliced alignment

In cDNA-to-genome alignment, we may want to distinguish introns from deletions in exons. We introduce openation 'N' to represent long skip on the reference sequence. Suppose the spliced alignment is:

spliced_alignment

```
REF:  AGCTAGCATCGTGTCGCCCGTCTAGCATACGCATGATCGACTGTCAGCTAGTCAGACTAGTCGATCGATGTG

READ:      GTGTAACCC..............................TCAGAATA
```

where '...' on the read sequence indicates intron. The CIGAR for this alignment is : 9M32N8M.

Multi-part alignment

One query sequence may be aligned to multiple places on the reference genome, either with or without overlaps. In SAM, we keep multiple hits as multiple alignment records. To avoid presenting the full query sequence multiple times for non-overlapping hits, we introduce operation 'H' to describe hard clipped alignment. Hard clipping (H) is similar to soft clipping (S). They are different in that hard clipped subsequence is not present in the alignment record. The example alignment in "clipped alignment" can also be represented with CIGAR: 3H8M1D6M4H, but in this case, the sequence stored in SAM is "GTGTAACCGACTAG", instead of "GGGGTGTAACCGACTAGGGGG" if soft clipping is in use.

Padded Alignment

Most sequence aligners only give the sequences inserted to the reference genome, but do not present how these inserted sequences are aligned against each other. Alignment with inserted sequences fully aligned is called padded alignment. Padded alignment is always produced by de novo assemblers and is important for an alignment viewer to display the alignment properly. To store padded alignment, we introduce operation 'P' which can be considered as a silent deletion from padded reference sequence. In the following example, GA on READ1 and A on READ2 are inserted to the reference. With unpadded CIGAR, we would not be able to distinguish the following padded multi-alignments:

padded_alignment

REF: CACGATCA**GACCGATACGTCCGA

READ1: CGATCAGAGACCGATA

READ2: ATCA*AGACCGATAC

READ3: GATCA**GACCG

The padded CIGAR are different:

READ1: 6M2I8M

READ2: 4M1P1I9M

READ3: 5M2P5M

Note that it is hard to convert unpadded CIGAR to padded one. Fully resolving the alignment between inserted sequences would essentially require a de novo assembler. However, it is easy vice versa. By simply removing all P operations we get the CIGAR without padding.

Alignments in color space Color alignments are stored as normal nucleotide alignments with additional tags describing the raw color sequences, qualities and color-specific properties.

Pileup Format

Pileup format is a text-based format for summarizing the base calls of aligned reads to a reference sequence. This format facilitates visual display of SNP/indel calling and alignment. It was first used by Tony Cox and Zemin Ning at the Wellcome Trust Sanger Institute, but became widely known through its implementation within the SAMtools software suite.

Format

Example

Sequence	Position	Reference Base	Read Count	Read Results	Quality
seq1	272	T	24	,.$.....,,.,.,...,,,.,..^+.	<<<+;<<<<<<<<<<<=<;<;7<&
seq1	273	T	23	,.....,,.,.,...,,,.,..A	<<<;<<<<<<<<<3<=<<<;<<+
seq1	274	T	23	,.$....,,.,.,...,,,.,...	7<7;<;<<<<<<<<<=<;<;<<6
seq1	275	A	23	,$....,,.,.,...,,,.,...^l.	<+;9*<<<<<<<<<=<<:;<<<<
seq1	276	G	22	...T,,.,.,...,,,.,....	33;+<<7=7<<7<&<<1;<<6<
seq1	277	T	22	....,,.,.,.C.,,,.,..G.	+7<;<<<<<<<&<=<<:;<<&<
seq1	278	G	23	....,,.,.,...,,,.,....^k.	%38*<<;<7<<7<=<<<;<<<<<
seq1	279	C	23	A..T,,.,.,...,,,.,.....	75&<<<<<<<<<=<<<9<<:<<<

Columns

Each line consists of 5 (or optionally 6) tab-separated columns:

1. Sequence identifier
2. Position in sequence (starting from 1)
3. Reference nucleotide at that position
4. Number of aligned reads covering that position (depth of coverage)
5. Bases at that position from aligned reads
6. Phred Quality of those bases, represented in ASCII with -33 offset (OPTIONAL)

Column 5: The Bases String

- . (dot) means a base that matched the reference on the forward strand
- , (comma) means a base that matched the reference on the reverse strand
- </> (less-/greater-than sign) denotes a reference skip. This occurs, for example, if a base in the reference genome is intronic and a read maps to two flanking exons. If quality scores are given in a sixth column, they refer to the quality of the read and not the specific base.
- AGTCN denotes a base that did not match the reference on the forward strand
- agtcn denotes a base that did not match the reference on the reverse strand
- A sequence matching the regular expression \+[0-9]+[ACGTNacgtn]+ denotes an insertion of one or more bases starting from the next position
- A sequence matching the regular expression -[0-9]+[ACGTNacgtn]+ denotes a deletion of one or more bases starting from the next position
- ^ (caret) marks the start of a read segment and the ASCII of the character following `^' minus 33 gives the mapping quality
- $ (dollar) marks the end of a read segment
- * (asterisk) is a placeholder for a deleted base in a multiple basepair deletion that was mentioned in a previous line by the -[0-9]+[ACGTNacgtn]+ notation

Column 6: The Base Quality String

This is an optional column. If present, the ASCII value of the character minus 33 gives the mapping Phred quality of each of the bases in the previous column 5. This is similar to quality encoding in the FASTQ format.

Stockholm Format

The "Stockholm" format is a system for marking up features in a multiple alignment. These mark-up annotations are preceded by a 'magic' label, of which there are four types. The Stockholm format is used by HMMER, Pfam, and Belvu.

The complete specification of the Stockholm format, is:

Header

The first line in the file must contain a format and version identifier, currently:

STOCKHOLM 1.0

Sequence Alignment

```
<seqname> <aligned sequence>
<seqname> <aligned sequence>
<seqname> <aligned sequence>
.
.
.
//
```

<seqname> stands for "sequence name", typically in the form "name/start-end" or just "name".

The "//" line indicates the end of the alignment.

Sequence letters may include any characters except whitespace. Gaps may be indicated by "." or "-".

Wrap-around alignments are allowed in principle, mainly for historical reasons, but are not used in e.g. Pfam. Wrapped alignments are discouraged since they are much harder to parse.

Alignment Mark-up

Mark-up lines may include any characters except whitespace. Use underscore ("_") instead of space.

```
#=GF <feature> <Generic per-File annotation, free text>
#=GC <feature> <Generic per-Column annotation, exactly 1 char per column>
#=GS <seqname> <feature> <Generic per-Sequence annotation, free text>
#=GR <seqname> <feature> <Generic per-Sequence AND per-Column markup,
exactly 1 char per column>
```

Magic or Recommended Features

#=GF

> For embedding trees:

```
#=GF NH <tree in New Hampshire eXtended format>


#=GF TN <Unique identifier for the next tree>
```

- A tree may be stored on multiple #=GF NH lines.

- If multiple trees are stored in the same file, each tree must be preceded by a #=GF TN line with a unique tree identifier. If only one tree is included, the #=GF TN line may be omitted.

#=GC

The same features as for #=GR with "_cons" appended, meaning "consensus". Example: "SS_cons".

#=GS

Pfam uses these features:

```
Feature                     Description

--------------------        -----------

AC <accession>              ACcession number

DE <freetext>               DEscription

DR <db>; <accession>;       Database Reference

OS <organism>               OrganiSm (species)

OC <clade>                  Organism Classification (clade, etc.)

LO <look>                   Look (Color, etc.)
```

#=GR

```
Feature       Description             Markup letters

-------       -----------             --------------

SS            Secondary Structure     [HGIEBTSCX]

SA            Surface Accessibility   [0-9X]

              (0=0%-10%; ...;      9=90%-100%)

TM            TransMembrane           [Mio]

PP            Posterior Probability   [0-9*]

              (0=0.00-0.05; 1=0.05-0.15; *=0.95-1.00)

LI            LIgand binding          [*]

AS            Active Site             [*]

pAS           AS - Pfam predicted     [*]

sAS           AS - from SwissProt     [*]

IN            INtron (in or after)    [0-2]
```

Do not use multiple lines with the same #=GR label. Only one unique feature assignment can be made for each sequence.

"X" in SA and SS means "residue with unknown structure".

In SS the letters are taken from DSSP: H=alpha-helix, G=3/10-helix, I=p-helix, E=extended strand, B=residue in isolated b-bridge, T=turn, S=bend, C=coil/loop.)

Recommended Placements

#=GF Above the alignment
#=GC Below the alignment
#=GS Above the alignment or just below the corresponding sequence
#=GR Just below the corresponding sequence

Size Limits

No size limits on any field.

However, a simple parser that uses fixed field sizes should work safely on Pfam alignments with these limits:

Line length: 10000.
<seqname>: 255.
<feature>: 255.

Example

```
# STOCKHOLM 1.0

#=GF ID CBS

#=GF AC PF00571

#=GF DE CBS domain

#=GF AU Bateman A

#=GF CC CBS domains are small intracellular modules mostly found

#=GF CC in 2 or four copies within a protein.

#=GF SQ 67

#=GS O31698/18-71 AC O31698

#=GS O83071/192-246 AC O83071

#=GS O83071/259-312 AC O83071

#=GS O31698/88-139 AC O31698

#=GS O31698/88-139 OS Bacillus subtilis

O83071/192-246          MTCRAQLIAVPRASSLAE..AIACAQKM....RVSRVPVYERS
```

```
#=GR O83071/192-246 SA 999887756453524252..55152525....36463774777

O83071/259-312          MQHVSAPVFVFECTRLAY..VQHKLRAH....SRAVAIVLDEY

#=GR O83071/259-312 SS  CCCCCHHHHHHHHHHHHH..EEEEEEEE....EEEEEEEEEEE

O31698/18-71            MIEADKVAHVQVGNNLEH..ALLVLTKT....GYTAIPVLDPS

#=GR O31698/18-71 SS    CCCHHHHHHHHHHHHHHHH..EEEEEEEE....EEEEEEEEHHH

O31698/88-139           EVMLTDIPRLHINDPIMK..GFGMVINN......GFVCVENDE

#=GR O31698/88-139 SS   CCCCCCCHHHHHHHHHHHH..HEEEEEEE....EEEEEEEEEEH

#=GC SS_cons            CCCCCHHHHHHHHHHHHHH..EEEEEEEE....EEEEEEEEEEH

O31699/88-139           EVMLTDIPRLHINDPIMK..GFGMVINN......GFVCVENDE

#=GR O31699/88-139 AS   _____*_____

#=GR_O31699/88-139_IN   _____1_____2_____0____
```

DNA and RNA Sequencing

Determining the precise order of nucleotides in a DNA molecule is called DNA sequencing. The investigation of the presence and quantity of RNA at a given moment in a biological sample is done using RNA sequencing. This chapter has been carefully written to provide an easy understanding of the varied aspects of DNA patterns, RNA and DNA sequencing.

DNA Sequencing

DNA sequencing is the determination of the precise sequence of nucleotides in a sample of DNA. Before the development of direct DNA sequencing methods, DNA sequencing was difficult and indirect. The DNA had to be converted to RNA, and limited RNA sequencing could be done by the existing cumbersome methods. Thus, only shorter DNA sequences could be determined by this method. Using this method, Walter Gilbert and Alan Maxam at Havard University determined that the Lac operator is a 27 bp long sequence.

The development of direct DNA sequencing techniques changed the scope of biological research. The evolution of DNA sequencing technology from plus-minus sequencing to pyro-sequencing within about 20 years parallels the progress in biology from molecular biology to genomics.

The development of DNA sequencing techniques with enhanced speed, sensitivity and through-put are of utmost importance for the study of biological systems. Sequence determination is most commonly performed using di-deoxy chain termination technology. Pyro-sequencing, a non-electrophoretic real- time bio-luminometric method for DNA sequencing has emerged as a state of the art sequencing technology.

This technology has the advantage of accuracy, ease of use, and high flexibility for different applications. Pyro-sequencing allows the analysis of genetic variations including SNPs, insertion/deletions and short repeats, as well as assessing RNA allelic imbalance, DNA methylation status and gene copy number.

Sanger's Method

The first DNA sequencing method devised by Sanger and Coulson in 1975 was called plus and minus sequencing that utilized E. coli DNA pol I and DNA polymerase from bacteriophage T4 with different limiting triphosphates. This technique had a low efficiency. Sanger and co-worker eventually invented a new method for DNA sequencing via enzymatic polymerization that basically revolutionized DNA sequencing technology.

The most popular method for doing this is called the dideoxy method or Sanger method (named after its inventor, Frederick Sanger, who was awarded the 1980 Nobel prize in chemistry [his second] for this achievement). Finding a single gene amid the vast stretches of DNA that make up the human genome – three billion base-pairs' worth – requires a set of powerful tools. These tools include genetic maps, physical maps and DNA sequence which is a detailed description of the order of the chemical building blocks, or bases, in a given stretch of DNA.

Scientists need to know the sequence of bases because it tells them the kind of genetic information that is carried in a particular segment of DNA. For example, they can use sequence information to determine which stretches of DNA contain genes, as well as to analyze those genes for changes in sequence, called mutations, that may cause disease.

The first methods for sequencing DNA were developed in the mid-1970s. At that time, scientists could sequence only a few base pairs per year, not nearly enough to sequence a single gene, much less the entire human genome. By the time the HGP began in 1990, only a few laboratories had managed to sequence a mere 100,000 bases, and the cost of sequencing remained very high. Since then, technological improvements and automation have increased speed and lowered cost to the point where individual genes can be sequenced routinely, and some labs can sequence well over 100 million bases per year.

DNA is synthesized from four deoxynucleotide triphosphates. The top formula shows one of them: deoxythymidine triphosphate. Each new nucleotide is added to the 3′ – OH group of the last nucleotide added.

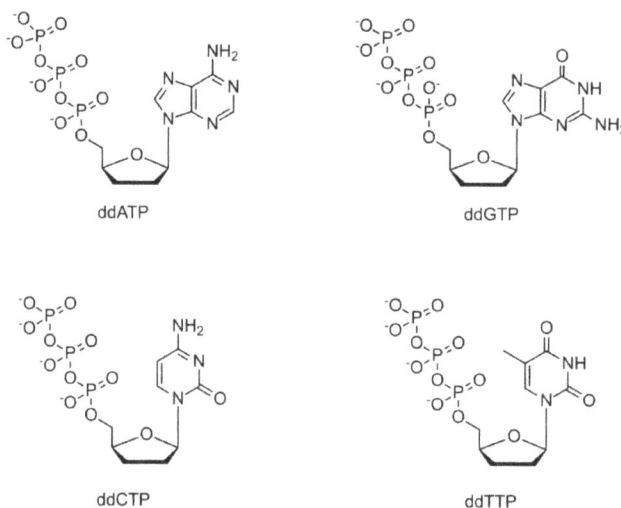

Structure of dideoxynucleotide

The dideoxy method gets its name from the critical role played by synthetic nucleotides that lack the -OH at the 3′ carbon atom. A dideoxynucleotide (dideoxythymidine triphosphate – ddTTP as shown here) can be added to the growing DNA strand. When it is added it stops chain elongation because there is no 3′ -OH for the next nucleotide to be attached. For this reason, the dideoxy method is also called the chain termination method.

The bottom formula shows the structure of azidothymidine (AZT), a drug used to treat AIDS. AZT (which is also called zidovudine) is taken up by cells where it is converted into the triphosphate. The reverse transcriptase of the human immunodeficiency virus (HIV) prefers AZT triphosphate to the normal nucleotide (dTTP). Because AZT has no 3′ -OH group, DNA synthesis by reverse transcriptase halts when AZT triphosphate is incorporated in the growing DNA strand. FortuWnately, the DNA polymerases of the host cell prefer dTTP, so side effects from the drug are not as severe as might have been predicted.

Procedure

The DNA to be sequenced is prepared as a single strand. This template DNA is mixed with the following:

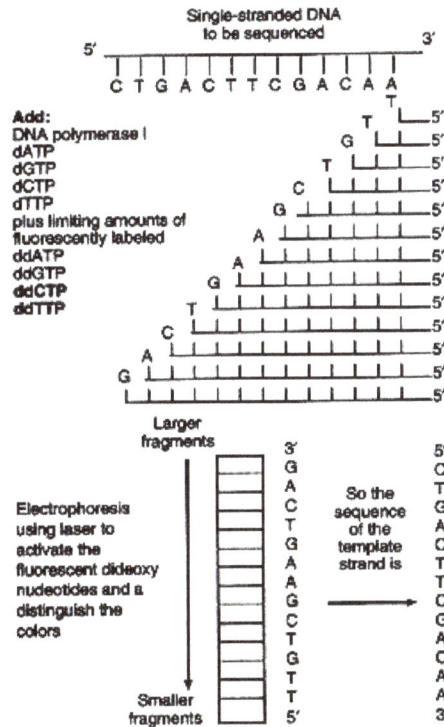

(a) A mixture of all four normal (deoxy) nucleotides in sample quantities

i. dATP

ii. dGTP

iii. dCTP

iv. dTTP

(b) A mixture of all four dideoxynucleotides, each present in limiting quantities and each labeled with a "tag" that fluoresces a different colour:

i. ddATP

ii. ddGTP

iii. ddCTP

iv. ddTTP

(c) DNA polymerase I

Because all four normal nucleotides are present, chain elongation proceeds normally until, by chance, DNA polymerase inserts a dideoxy nucleotide instead of the normal deoxynucleotide. If the ratio of normal nucleotide to the dideoxy versions is high enough, some DNA strands will succeed in adding several hundred nucleotides before insertion of the dideoxy version halts the process.

At the end of the incubation period, the fragments are separated by length from longest to shortest. The resolution is so good that a difference of one nucleotide is enough to separate that strand from the next shorter and next longer strand. Each of the four dideoxynucleotides fluoresces a different color when illuminated by a laser beam and an automatic scanner provides a printout of the sequence.

Next-generation Sequencing

The most recent set of DNA sequencing technologies are collectively referred to as next-generation sequencing.

There are a variety of next-generation sequencing techniques that use different technologies. However, most share a common set of features that distinguish them from Sanger sequencing:

- Highly parallel: many sequencing reactions take place at the same time

- Micro scale: reactions are tiny and many can be done at once on a chip

- Fast: because reactions are done in parallel, results are ready much faster

- Low-cost: sequencing a genome is cheaper than with Sanger sequencing

- Shorter length: reads typically range from 50 -70 nucleotides in length

Conceptually, next-generation sequencing is kind of like running a very large number of tiny Sanger sequencing reactions in parallel. Thanks to this parallelization and small scale, large quantities of DNA can be sequenced much more quickly and cheaply with next-generation methods than with Sanger sequencing. For example, in 2001, the cost of sequencing a humangenome was almost $100 million. In 2015, it was just $1245.

Why does fast and inexpensive sequencing matter? The ability to routinely sequence genomes opens new possibilities for biology research and biomedical applications. For example, low-cost sequencing is a step towards personalized medicine – that is, medical treatment tailored to an individual's needs, based on the gene variants in his or her genome.

Application

DNA sequencing provides the means to know how nucleotide bases are arranged in a piece of DNA. Several methods have been developed for this process. These have four key steps. In the first instance DNA is removed from the cell. This can be done either mechanically or chemically. The second phase involves breaking up the DNA and inserting its pieces into vectors, cells that indefinitely self-replicate, for cloning. In the third phase the DNA clones are placed with a dye-labelled primer (a short stretch of DNA that promotes replication) into a thermal cycler, a machine which automatically raises and lowers the temperature to catalyze replication. The final phase consists of electrophoresis, whereby the DNA segments are placed in a gel and subjected to an electrical current which moves them. Originally the gel was placed on a slab, but today it is inserted into a very thin glass tube known as a capillary. When subjected to an electrical current the smaller nucleotides in the DNA move faster than the larger ones. Electrophoresis thus helps sort out the DNA fragments by their size. The different nucleotide bases in the DNA fragments are identified by their dyes which are activated when they pass through a laser beam. All the information is fed into a computer and the DNA sequence displayed on a screen for analysis.

The method developed by Sanger was pivotal to the international Human Genome Project. Costing over US$3 billion and taking 13 years to complete, this project provided the first complete Human DNA sequence in 2003. Data from the project provided the first means to map out the genetic mutations that underlie specific genetic diseases. It also opened up a path to more personalized medicine, enabling scientists to examine the extent to which a patient's response to a drug is determined by their genetic profile. The genetic profile of a patient's tumor, for example, can now be used to work out what is the most effective treatment for an individual. It is also hoped that in the future that knowing the sequence of a person's genome will help work out a person's predisposition to certain diseases, such as heart disease, cancer and type II diabetes, which could pave the way to better preventative care.

Data from the Human Genome Project has also helped fuel the development of gene therapy, a type of treatment designed to replace defective genes in certain genetic disorders. In addition, it has provided a means to design drugs that can target specific genes that cause disease.

Beyond medicine, DNA sequencing is now used for genetic testing for paternity and other family relationships. It also helps identify crime suspects and victims involved in catastrophes. The technique is also vital to detecting bacteria and other organisms that may pollute air, water, soil and food. In addition the method is important to the study of the evolution of different population groups and their migratory patterns as well as determining pedigree for seed or livestock.

DNA Nanoball Sequencing

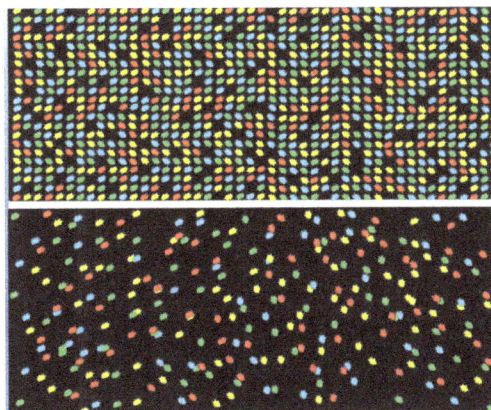

Nanoball Sequencing is the advanced form of DNA sequencing used to determine the entire gene sequence of an organism. The DNA Nanoball sequencing technique uses the Rolling Circle model of bacterial replication to convert the DNA of genes into Nanoballs. Fluorescent probes are used to bind with the complementary sequences, and then these markers are used to locate the sequence in the DNA template. DNA Nanoball technique allows sequencing large number of DNA segments very fast at low cost. DNA Nanoball sequencing is a very good technique for identifying Mutations in genes and Mendelian disorders.

Protocol of Nanoball Sequencing Involves

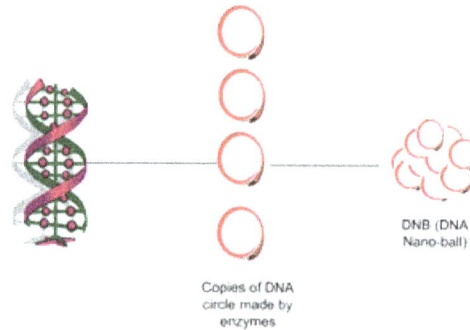

DNB (DNA Nano-ball)

Copies of DNA circle made by enzymes

- Isolation of the desired DNA and converting it into small segments having 400-500 base pairs (bp).

- Ligating the Adapter sequences to the DNA sequence and converting the sequence into circular fragments.

- Using the Rolling Circle replication, these circular fragments are allowed to replicate and produce single stranded circular fragments.

- The single stranded circular fragments thus produced are then allowed to concatenate head to tail order to produce a DNA Nanoball.

- The Nanoballs are then allowed to adsorb on to a Flow cell.

- Fluorescent probes are then allowed to ligate with the specific nucleotide sequence.

- The color of the fluorescence at each location is then recorded using a high resolution camera.

- Using the Bioinformatics techniques the data is then analyzed.

- The genomic data is then assembled and the sequence is identified.

Extraction of DNA

DNA from cells is extracted through cell lysis using tissue fragmentation. The DNA thus obtained is mega base pair long which is sonicated using Ultrasound to break it into segments at random intervals. The fragments are then separated using PolyAcrylamide Gel Electrophoresis (PAGE). After the PAGE separation, the separated fragments are purified by Gel extraction resulting in small length segments having 400 to 500 base pairs.

Creation of Circular DNA

After the DNA fragmentation, Adapter DNA sequences are allowed to attach with the DNA segment of unknown sequence. The DNA is then amplified using PCR. The DNA is then modified to create single stranded ends which can be joined together to form a circular shaped DNA. Restriction endonuclease is used to cleave the DNA strand at 13 bp site to form the linear segment. During the process 4 adapter molecules are allowed to bind with the DNA segment. The Circular DNA template now contains 4 adapters.

Rolling Circle Replication

The full circular DNA is then amplified into a long string of DNA through Rolling circle replication using the enzyme Phi29 DNA Polymerase. The newly synthesized DNA is then separated from the circular DNA template resulting in a long DNA with copies of the circular DNA.

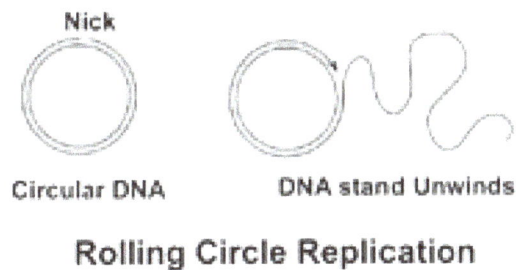

Nick

Circular DNA   DNA stand Unwinds

Rolling Circle Replication

Formation of Nanoball

The 4 adapter sequences are still in the DNA that contains Palindromic sequences. The adapter sequences hybridize so that the single strand to fold itself into a tight ball of DNA with 300 nanometers across. The Nanoballs remain separated each other preventing the tangling in the single stranded DNA. To get the DNA sequence, the DNA Nanoballs are attached to a Flow cell which is 25mm by 75 mm silicon wafer. The Silicon wafer is coated with Silicon dioxide, titanium, hexamethyldisilazane and a Photoresist material. The Nanoballs in the Flow cell are allowed to bind with Aminosilane to create high density DNA Nanoballs.

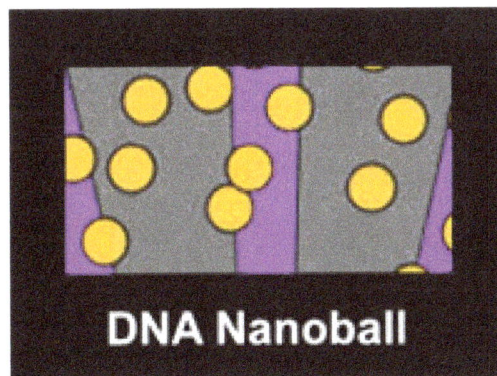

DNA Nanoball

Imaging and Data Analysis

The order of DNA sequence is determined after arrayed on the Flow cell. Oligonucleotide complementary to one of the adapter is added along with DNA ligase enzyme. The positions of different Nitrogen bases in the DNA probe is shown by the Fluorophore attached.

- Adenine contains Red fluorophore

- Cytosine with Yellow fluorophore

- Guanine with Green fluorophore

- Thymine with Blue fluorophore

Only the Probe with complementary nucleotide will binds with the corresponding nucleotide in the DNA. The Non binding probes will be washed away. By observing the Fluorescence, it is easy to locate the position and type of base in the DNA. After identification, the probe is removed from the DNA and another probe is added to locate other nucleotides.

The Flow cell is then imaged to identify the base attached to the DNA Nanoball. For this, an Arc lamp is used. It illuminates the Nanoballs with specific wavelength light. A high resolution CCD (Charge Coupled Device) camera collects the wavelength of florescence from the Nanoball. The image is then processed and the computer records the position of base depending on the color of fluorescence.

Advantages of DNA Nanoball Sequencing Include

- It uses High density array so that high concentration of DNA can be used.

- Sequencing reaction is not progressive so that the new probes can be added after removing the probe already given.

- Accurate amplification using High fidelity Phi 29 DNA polymerase.

Transmission Electron Microscopy DNA Sequencing

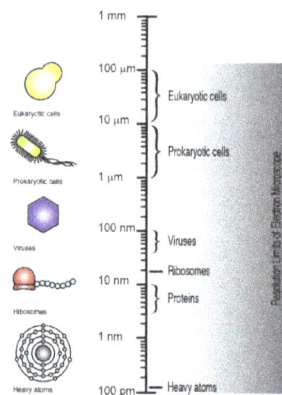

The electron microscope can achieve a resolution of up to 100 picometers, allowing eukaryotic cells, prokaryotic cells, viruses, ribosomes, and even single atoms to be visualized.

Transmission electron microscopy DNA sequencing is a single-molecule sequencing technology that uses transmission electron microscopy techniques. The method was conceived and developed in the 1960s and 70s, but lost favor when the extent of damage to the sample was recognized.

In order for DNA to be clearly visualized under an electron microscope, it must be labeled with heavy atoms. In addition, specialized imaging techniques and aberration corrected optics are beneficial for obtaining the resolution required to image the labeled DNA molecule. In theory, transmission electron microscopy DNA sequencing could provide extremely long read lengths, but the issue of electron beam damage may still remain and the technology has not yet been commercially developed.

Principle

The electron microscope has the capacity to obtain a resolution of up to 100 pm, whereby microscopic biomolecules and structures such as viruses, ribosomes, proteins, lipids, small molecules and even single atoms can be observed.

Although DNA is visible when observed with the electron microscope, the resolution of the image obtained is not high enough to allow for deciphering the sequence of the individual bases, *i.e.*, DNA sequencing. However, upon differential labeling of the DNA bases with heavy atoms or metals, it is possible to both visualize and distinguish between the individual bases. Therefore, electron microscopy in conjunction with differential heavy atom DNA labeling could be used to directly image the DNA in order to determine its sequence.

Workflow

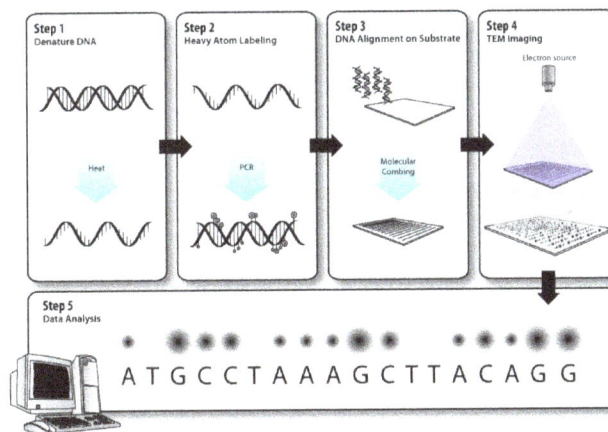

Workflow of transmission electron microscopy DNA sequencing

Step 1 – DNA Denaturation

As in a standard polymerase chain reaction (PCR), the double stranded DNA molecules to be sequenced must be denatured before the second strand can be synthesized with labeled nucleotides.

Step 2 – Heavy Atom Labeling

The elements that make up biological molecules (C, H, N, O, P, S) are too light (low atomic number, Z) to be clearly visualized as individual atoms by transmission electron microscopy. To circumvent this problem, the DNA bases can be labeled with heavier atoms (higher Z). Each nucleotide is tagged with a characteristic heavy label, so that they can be distinguished in the transmission electron micrograph.

- ZS Genetics proposes using three heavy labels: bromine (Z=35), iodine (Z=53), and tri-chloromethane (total Z=63). These would appear as differential dark and light spots on the micrograph, and the fourth DNA base would remain unlabeled.

- Halcyon Molecular, in collaboration with the Toste group, proposes that purine and pyrimidine bases can be functionalized with platinum diamine or osmium tetraoxide bipyridine, respectively. Heavy metal atoms such as osmium (Z=76), iridium (Z=77), gold (Z=79), or uranium (Z=92) can then form metal-metal bonds with these functional groups to label the individual bases.

Step 3 – DNA Alignment on Substrate

The DNA molecules must be stretched out on a thin, solid substrate so that order of the labeled bases will be clearly visible on the electron micrograph. Molecular combing is a technique that utilizes the force of a receding air-water interface to extend DNA molecules, leaving them irreversibly bound to a silane layer once dry. This is one means by which alignment of the DNA on a solid substrate may be achieved.

Step 4 – Transmission Electron Microscopy Imaging

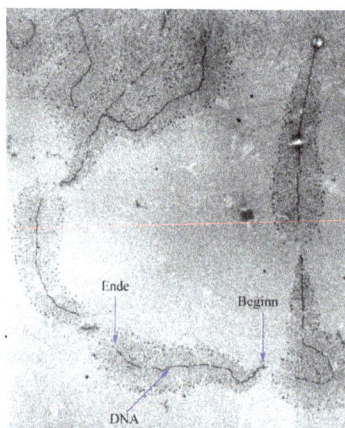

Electron microscopy image of DNA: ribosomal transcription units of *Chrironumus pallidivitatus*.

Transmission electron microscopy (TEM) produces high magnification, high resolution images by passing a beam of electrons through a very thin sample. Whereas atomic resolution has been demonstrated with conventional TEM, further improvement in spatial resolution requires correcting the spherical and chromatic aberrations of the microscope lenses. This has only been possible in scanning transmission electron microscopy where the image is obtained by scanning the object with a finely focused electron beam, in a way similar to a cathode ray tube. However, the achieved improvement in resolution comes together with irradiation of the studied object by much higher beam intensities, the concomitant sample damage and the associated imaging artefacts. Different imaging techniques are applied depending on whether the sample contains heavy or light atoms:

- Annular dark-field imaging measures the scattering of electrons as they deflect off the nuclei of the atoms in the transmission electron microscopy sample. This is best suited to samples containing heavy atoms, as they cause more scattering of electrons. The technique has been used to image atoms as light as boron, nitrogen, and carbon; however, the signal is very weak for such light atoms. If annular dark-field microscopy is put to use for transmission electron microscopy DNA sequencing, it will certainly be necessary to label the DNA bases with heavy atoms so that a strong signal can be detected.

- Annular bright-field imaging detects electrons transmitted directly through the sample, and measures the wave interference produced by their interactions with the atomic nuclei. This technique can detect light atoms with greater sensitivity than annular dark-field imaging methods. In fact, oxygen, nitrogen, lithium, and hydrogen in crystalline solids have been imaged using annular bright-field electron microscopy. Thus, it is theoretically possible to obtain direct images of the atoms in the DNA chain; however, the structure of DNA is much less geometric than crystalline solids, so direct imaging without prior labeling may not be achievable.

Step 5 – Data Analysis

Dark and bright spots on the electron micrograph, corresponding to the differentially labeled DNA bases, are analyzed by computer software.

Applications

Transmission electron microscopy DNA sequencing is not yet commercially available, but the long read lengths that this technology may one day provide will make it useful in a variety of contexts.

De novo Genome Assembly

When sequencing a genome, it must be broken down into pieces that are short enough to be sequenced in a single read. These reads must then be put back together like a jigsaw puzzle by aligning the regions that overlap between reads; this process is called *de novo* genome assembly. The longer the read length that a sequencing platform provides, the longer the overlapping regions, and the easier it is to assemble the genome. From a computational perspective, microfluidic Sanger sequencing is still the most effective way to sequence and assemble genomes for which no reference genome sequence exists. The relatively long read lengths provide substantial overlap between individual sequencing reads, which allows for greater statistical confidence in the assembly. In addition, long Sanger reads are able to span most regions of repetitive DNA sequence which otherwise confound sequence assembly by causing false alignments. However, *de novo* genome assembly by Sanger sequencing is extremely expensive and time consuming. Second generation sequencing technologies, while less expensive, are generally unfit for *de novo* genome assembly due to short read lengths. In general, third generation sequencing technologies, including transmission electron microscopy DNA sequencing, aim to improve read length while maintaining low sequencing cost. Thus, as third generation sequencing technologies improve, rapid and inexpensive *de novo* genome assembly will become a reality.

Full Haplotypes

A haplotype is a series of linked alleles that are inherited together on a single chromosome. DNA sequencing can be used to genotype all of the single nucleotide polymorphisms (SNPs) that constitute a haplotype. However, short DNA sequencing reads often cannot be phased; that is, heterozygous variants cannot be confidently assigned to the correct haplotype. In fact, haplotyping with short read DNA sequencing data requires very high coverage (average >50x coverage of each DNA base) to accurately identify SNPs, as well as additional sequence data from the parents so that Mendelian transmission can be used to estimate the haplotypes. Sequencing technologies that generate long reads, including transmission electron microscopy DNA sequencing, can capture entire haploblocks in a single read. That is, haplotypes are not broken up among multiple reads, and the genetically linked alleles remain together in the sequencing data. Therefore, long reads make haplotyping easier and more accurate, which is beneficial to the field of population genetics.

Copy Number Variants

Genes are normally present in two copies in the diploid human genome; genes that deviate from this standard copy number are referred to as copy number variants (CNVs). Copy number variation can be benign (these are usually common variants, called copy number polymorphisms) or pathogenic. CNVs are detected by fluorescence in situ hybridization (FISH) or comparative genomic hybridization (CGH). To detect the specific breakpoints at which a deletion occurs, or to detect genomic lesions introduced by a duplication or amplification event, CGH can be performed using a tiling array (array CGH), or the variant region can be sequenced. Long sequencing reads are especially useful for analyzing duplications or amplifications, as it is possible to analyze the orientation of the amplified segments if they are captured in a single sequencing read.

Cancer

Cancer genomics, or oncogenomics, is an emerging field in which high-throughput, second generation DNA sequencing technology is being applied to sequence entire cancer genomes. Analyzing this short read sequencing data encompasses all of the problems associated with *de novo* genome assembly using short read data. Furthermore, cancer genomes are often aneuploid. These aberrations, which are essentially large scale copy number variants, can be analyzed by second-generation sequencing technologies using read frequency to estimate the copy number. Longer reads would, however, provide a more accurate picture of copy number, orientation of amplified regions, and SNPs present in cancer genomes.

Microbiome Sequencing

The microbiome refers the total collection of microbes present in a microenvironment and their respective genomes. For example, an estimated 100 trillion microbial cells colonize the human body at any given time. The human microbiome is of particular interest, as these commensal bacteria are important for human health and immunity. Most of the Earth's bacterial genomes have not yet been sequenced; undertaking a microbiome sequencing project would require extensive *de novo* genome assembly, a prospect which is daunting with short read DNA sequencing technologies. Longer reads would greatly facilitate the assembly of new microbial genomes.

Strengths and Weaknesses

Compared to other second- and third-generation DNA sequencing technologies, transmission electron microscopy DNA sequencing has a number of potential key strengths and weaknesses, which will ultimately determine its usefulness and prominence as a future DNA sequencing technology.

Strengths

- Longer read lengths: ZS Genetics has estimated potential read lengths of transmission electron microscopy DNA sequencing to be 10,000 to 20,000 base pairs with a rate of 1.7 billion base pairs per day. Such long read lengths would allow easier *de novo* genome assembly and direct detection of haplotypes, among other applications.

- Lower cost: Transmission electron microscopy DNA sequencing is estimated to cost just US$5,000-US$10,000 per human genome, compared to the more expensive second-generation DNA sequencing alternatives.

- No dephasing: Dephasing of the DNA strands due to loss in synchronicity during synthesis is a major problem of second-generation sequencing technologies. For transmission electron microscopy DNA sequencing and several other third-generation sequencing technologies, sychronization of the reads is unnecessary as only one molecule is being read at a time.

- Shorter turnaround time: The capacity to read native fragments of DNA renders complex template preparation an unnecessary step in the general workflow of whole genome sequencing. Consequently, shorter turnaround times are possible.

Weaknesses

- High capital cost: A transmission electron microscope with sufficient resolution required for transmission electron microscopy DNA sequencing costs approximately US$1,000,000, therefore pursuing DNA sequencing by this method requires a substantial investment.

- Technically challenging: Selective heavy atom labeling and attaching and straightening the labeled DNA to a substrate are a serious technical challenge. Further, the DNA sample should be stable to the high vacuum of electron microscope and irradiation by a focused beam of high-energy electrons.

- Potential PCR bias and artefacts: Although PCR is only being utilized in transmission electron microscopy DNA sequencing as a means to label the DNA strand with heavy atoms or metals, there could be the possibility of introducing bias in template representation or errors during the single amplification.

Comparison to other Sequencing Technologies

Many non-Sanger second- and third-generation DNA sequencing technologies have been or are currently being developed with the common aim of increasing throughput and decreasing cost such that personalized genetic medicine can be fully realized.

Both the US$10 million Archon X Prize for Genomics supported by the X Prize Foundation (Santa Monica, CA, USA) and the US$70 million in grant awards supported by the National Human

Genome Research Institute of the National Institutes of Health (NIH-NHGRI) are fueling the rapid burst of research activity in the development of new DNA sequencing technologies.

Since different approaches, techniques, and strategies are what define each DNA sequencing technology, each has its own strengths and weaknesses. Comparison of important parameters between various second- and third-generation DNA sequencing technologies are presented in Table.

Table. Second- and third-generation DNA sequencing platforms						
Platform	Generation	Read length (bp)	Accuracy	Cost per human genome (US$)	Cost of instrument (US$)	Run time (h/Gbp)
Massively parallel pyrosequencing by synthesis	Second	400–500	Q20 read length of 40 bases (99% at 400 bases and higher for prior bases)	1,000,000	500,000	75
Sequencing by synthesis	Second	2×75	Base call with Q30 (>70%)	60,000	450,000	56
Bead-based massively parallel clonal ligation based sequencing	Second	100	99.94%	60,000	591,000	42
Massively parallel single-molecule sequencing by synthesis	Third	30–35	99.995% at >20×coverage (raw error rate: ≤ 5%)	70,000	1,350,000	~12
Single molecule, real time sequencing by synthesis	Third	1000–1500	99.3% at 15×coverage (error rate of a single read: 15–20%)	–	–	<1
Nanopore sequencing	Third	Potentially unlimited?	--	--	--	>20
Transmission electron microscopy single-molecule sequencing (ZS Genetics, Halcyon Molecular)	Third	Potentially unlimited?	--	~10,000	~1,000,000	~14

Ion Semiconductor Sequencing

Ion semiconductor sequencing is a method of DNA sequencing based on the detection of hydrogen ions that are released during the polymerization of DNA. This is a method of "sequencing by synthesis", during which a complementary strand is built based on the sequence of a template strand.

A Ion Proton semiconductor sequencer

A microwell containing a template DNA strand to be sequenced is flooded with a single species of deoxyribonucleotide triphosphate (dNTP). If the introduced dNTP is complementary to the leading template nucleotide, it is incorporated into the growing complementary strand. This causes the release of a hydrogen ion that triggers an ISFET ion sensor, which indicates that a reaction has occurred. If homopolymer repeats are present in the template sequence, multiple dNTP molecules will be incorporated in a single cycle. This leads to a corresponding number of released hydrogens and a proportionally higher electronic signal.

This technology differs from other sequencing technologies in that no modified nucleotides or optics are used. Ion semiconductor sequencing may also be referred to as Ion Torrent sequencing, pH-mediated sequencing, silicon sequencing, or semiconductor sequencing.

Technology

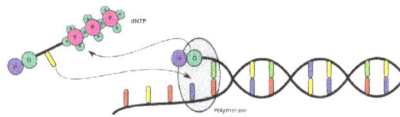

Polymerase integrates a nucleotide.

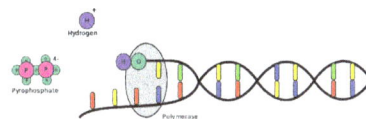

Hydrogen and pyrophosphate are released.

The incorporation of deoxyribonucleotide Triphosphate into a growing DNA strand causes the release of hydrogen and pyrophosphate.

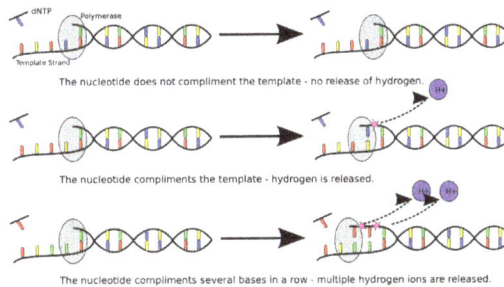

The release of hydrogen ions indicate if zero, one or more nucleotides were incorporated.

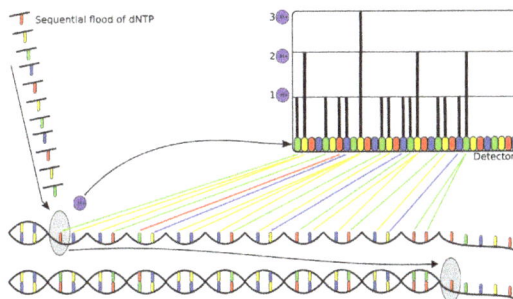

Released hydrogens ions are detected by an ion sensor. Multiple incorporations lead to a corresponding number of released hydrogens and intensity of signal.

Sequencing Chemistry

In nature, the incorporation of a deoxyribonucleoside triphosphate (dNTP) into a growing DNA strand involves the formation of a covalent bond and the release of pyrophosphate and a positively charged hydrogen ion. A dNTP will only be incorporated if it is complementary to the leading unpaired template nucleotide. Ion semiconductor sequencing exploits these facts by determining if a hydrogen ion is released upon providing a single species of dNTP to the reaction.

Microwells on a semiconductor chip that each contain many copies of one single-stranded template DNA molecule to be sequenced and DNA polymerase are sequentially flooded with unmodified A, C, G or T dNTP. If an introduced dNTP is complementary to the next unpaired nucleotide on the template strand it is incorporated into the growing complementary strand by the DNA polymerase. If the introduced dNTP is not complementary there is no incorporation and no biochemical reaction. The hydrogen ion that is released in the reaction changes the pH of the solution, which is detected by an ISFET. The unattached dNTP molecules are washed out before the next cycle when a different dNTP species is introduced.

Signal Detection

Beneath the layer of microwells is an ion sensitive layer, below which is an ISFET ion sensor. All layers are contained within a CMOS semiconductor chip, similar to that used in the electronics industry.

Each chip contains an array of microwells with corresponding ISFET detectors. Each released hydrogen ion then triggers the ISFET ion sensor. The series of electrical pulses transmitted from the chip to a computer is translated into a DNA sequence, with no intermediate signal conversion required. Because nucleotide incorporation events are measured directly by electronics, the use of labeled nucleotides and optical measurements are avoided. Signal processing and DNA assembly can then be carried out in software.

Sequencing Characteristics

The per base accuracy achieved in house by Ion Torrent on the Ion Torrent Ion semiconductor sequencer as of February 2011 was 99.6% based on 50 base reads, with 100 Mb per run. The read-length as of February 2011 was 100 base pairs. The accuracy for homopolymer repeats of 5 repeats in length was 98%. Later releases show a read length of 400 base pairs It should be noted that these figures have not yet been independently verified outside of the company.

Strengths

The major benefits of ion semiconductor sequencing are rapid sequencing speed and low upfront and operating costs. This has been enabled by the avoidance of modified nucleotides and optical measurements.

Because the system records natural polymerase-mediated nucleotide incorporation events, sequencing can occur in real-time. In reality, the sequencing rate is limited by the cycling of substrate nucleotides through the system. Ion Torrent Systems Inc., the developer of the technology, claims that each incorporation measurement takes 4 seconds and each run takes about one hour, during which 100-200 nucleotides are sequenced. If the semiconductor chips are improved (as predicted by Moore's law), the number of reads per chip (and therefore per run) should increase.

The cost of acquiring a pH-mediated sequencer from Ion Torrent Systems Inc. at time of launch was priced at around $50,000 USD, excluding sample preparation equipment and a server for data analysis. The cost per run is also significantly lower than that of alternative automated sequencing methods, at roughly $1,000.

Limitations

If homopolymer repeats of the same nucleotide (e.g. TTTTT) are present on the template strand (strand to be sequenced) then multiple introduced nucleotides are incorporated and more hydrogen ions are released in a single cycle. This results in a greater pH change and a proportionally greater electronic signal. This is a limitation of the system in that it is difficult to enumerate long repeats. This limitation is shared by other techniques that detect single nucleotide additions such as pyrosequencing. Signals generated from a high repeat number are difficult to differentiate from repeats of a similar but different number; *e.g.*, homorepeats of length 7 are difficult to differentiate from those of length 8.

Another limitation of this system is the short read length compared to other sequencing methods such as Sanger sequencing or pyrosequencing. Longer read lengths are beneficial for *de novo* genome assembly. Ion Torrent semiconductor sequencers produce an average read length of approximately 400 nucleotides per read.

The throughput is currently lower than that of other high-throughput sequencing technologies, although the developers hope to change this by increasing the density of the chip.

Application

The developers of Ion Torrent semiconductor sequencing have marketed it as a rapid, compact and economical sequencer that can be utilized in a large number of laboratories as a bench top machine. The company hopes that their system will take sequencing outside of specialized centers and into the reach of hospitals and smaller laboratories.

Due to the ability of alternative sequencing methods to achieve a greater read length (and therefore being more suited to whole genome analysis) this technology may be best suited to small scale applications such as microbial genome sequencing, microbial transcriptome sequencing, targeted sequencing, amplicon sequencing, or for quality testing of sequencing libraries.

Sequencing by Hybridization

Ed Southern's sequencing by hybridization technique relies on detection of specific DNA sequences using hybridization of complementary probes. It utilizes a large number of short nested oligonucleotides immobilized on a solid support to which the labeled sequencing template is hybridized. The target sequence is deduced by computer analysis of the hybridization pattern of the sample DNA.

DNA sequence can also be analyzed by sequencing by synthesis. Sequencing by hybridization makes use of a universal DNA microarray, which harbors all nucleotides of length k (called "k-words", or simply words when k is clear). These oligonucleotides are hybridize to an unknown DNA fragment, whose sequence one would like to determine.

Under ideal conditions, this target molecule will hybridize to all words whose Watson-Crick complements occur somewhere along its sequence. Thus, in principle, one would determine in a single microarray reaction the set of all k-long substrings of the target and try to infer the sequence from those data.

The average length of a uniquely resconstructible sequence using an 8-mer array is <200 bases, far below a single read length on commercial gel-lane machine. The main weakness of sequencing by hybridization is ambiguous solutions-when several sequences have the same spectrum; there is no way to determine the true sequence.

Single Molecule Real Time Sequencing

In an effort to overcome inherent challenges in the field of genomics, we sought to develop novel technology that pushed the boundaries of sequencing. The result, PacBio long-read sequencing enabled by SMRT Sequencing technology, harnesses the natural process of DNA replication and enables real-time observation of DNA synthesis. With this unique technology, we equip innovative scientists and deliver the results needed to drive genetic discovery.

SMRT Sequencing is built upon two key innovations: zero-mode waveguides (ZMWs) and phospholinked nucleotides. ZMWs allow light to illuminate only the bottom of a well in which a DNA polymerase/template complex is immobilized. Phospholinked nucleotides allow observation of the immobilized complex as the DNA polymerase produces a completely natural DNA strand.

Smrt Sequencing Technology and Terminology

Before SMRT sequencing, a library needs to be prepared from double stranded DNA input material. Typically this often requires five or more micrograms of DNA which can limit some applications. The library preparation consists of simply ligating hairpin adapters onto DNA molecules, thereby circularizing them into a construct termed a SMRTbell. Next, a primer and a polymerase are annealed to the adapter whereupon the library is loaded on a SMRT Cell containing 150 000 nanoscale observation chambers (Zero Mode Waveguides (ZMWs)) for the RSII system and up to a million on the newer Sequel platform. The polymerase bound SMRTbells are then loaded into the ZMWs. Ideally as many ZMWs should be loaded with exactly one SMRTbell as possible to maximise throughput and read lengths. For a good run, this is around one third to one half of the ZMWs per SMRT cell. Hence a SMRT cell typically produces ~55 000 reads for the RSII system and 365 000 reads for the Sequel system. The actual sequencing reaction occurs within each ZMW, whose small diameter only permits the smallest available volume for light detection. The polymerase within each ZMW incorporates fluorescently labeled nucleotides, emitting a fluorescent signal that is recorded by a camera in real-time. These signals are converted to long sequences termed continuous long reads (CLR), linear reads, or polymerase reads. For a short insert library, the circular structure of the molecule results in the insert sequence being covered multiple times by the CLR. Each pass of an original strand is termed a subread. In addition, all subreads from the same molecule can be combined into one highly accurate consensus sequence termed a circular consensus sequence (CCS) or reads-of-insert (ROI). These two terms are often used interchangeably, but by definition the difference is CCS requires two full sequencing passes of the insert whereas ROI can be defined starting from even a partial pass.

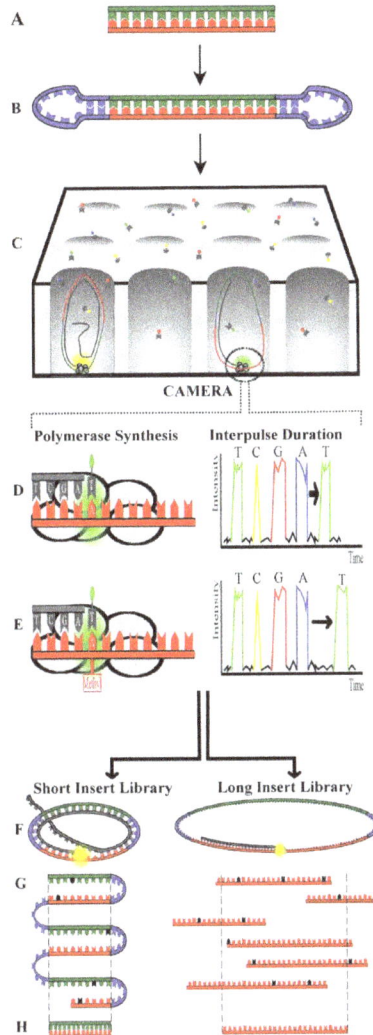

Overview of SMRT Sequencing Technology. Sequencing starts with preparing a library from double stranded DNA (A) to which hairpin adapters are ligated (B). This library is thereafter loaded onto a SMRT Cell made up of nanoscale observation chambers (Zero Mode Waveguides (ZMWs)). The DNA molecules in the library will be pulled to the bottom of the ZMW where the polymerase will incorporate fluorescently labelled nucleotides (C). Note that not all ZMWs will contain a DNA molecule because the library is loaded by diffusion. The fluorescence emitted by the nucleotides is recorded by a camera in real-time. Hence, not only the fluorescence color can be registered, but also the time between nucleotide incorporation which is called the interpulse duration (IPD) (D, right panel). When a sequencing polymerase encounters nucleotides on the DNA strand containing an (epigenetic) modification, like for example a 6-methyl adenosine modification (E, left panel), then the IPD will be delayed (E, right panel) compared to non-methylated DNA (D, right panel). Due to the circular structure of the library, a short insert will be covered multiple times by the continuous long read (CLR). Each pass of the original DNA molecule is termed a subread, which can be combined into one highly accurate consensus sequence termed a circular consensus sequence (CCS) or reads-of-insert (ROI) (F–H, left panel). Though SMRT sequencing always uses a circular template, long insert libraries typically only have a single pass and hence generate a linear sequence with single pass error rates (black nucleotides) (FG, right panel). Afterwards, overlapping single passes can be combined into one consensus sequence of high quality (H, right panel). Overall, CCS reads have the advantage of being very accurate while single passes stand out for their long read lengths (>20 kb).

Extending Read Lengths

So-called next-generation technologies for sequencing DNA are penetrating every aspect of biology thanks to the immense amount of information that is encoded within nucleic acid sequences. However, today's next-generation sequencing technologies, such as Illumina, 454 and Ion Torrent, have several significant limitations, especially short read lengths and amplification biases, that restrict our ability to fully sequence genomes. Unfortunately, with the rise of next-generation sequencing, even less emphasis is being placed on trying to understand at the biological and biochemical levels just what functions newly discovered genes have and how those functions allow an organism to work, which is surely why we are sequencing DNA in the first place. Now a new technology, SMRT sequencing from Pacific Biosciences, has been developed that not only produces considerably longer and highly accurate DNA sequences from individual unamplified molecules, but can also show where methylated bases occur (and thereby provide functional information about the DNA methyltransferases encoded by the genome).

SMRT sequencing is a sequencing-by-synthesis technology based on real-time imaging of fluorescently tagged nucleotides as they are synthesized along individual DNA template molecules. Because the technology uses a DNA polymerase to drive the reaction, and because it images single molecules, there is no degradation of signal over time. Instead, the sequencing reaction ends when the template and polymerase dissociate. As a result, instead of the uniform read length seen with other technologies, the read lengths have an approximately log-normal distribution with a long tail. The average read length from the current PacBio RS instrument is about 3,000 bp, but some reads may be 20,000 bp or longer. This is roughly 30 to 200 times longer than the read length from a next-generation sequencing instrument, and more than a four-fold improvement since the original release of the instrument two years ago. It is notable that the recently announced PacBio RS II platform claims to have a further four-fold improvement, with twice the mean read length and twice the throughput of the current machine.

Applications of SMRT Sequencing

The SMRT approach to sequencing has several advantages. First, consider the impact of the longer reads, especially for *de novo* assemblies of novel genomes. While typical next-generation sequencing can provide abundant coverage of a genome, the short read lengths and amplification biases of those technologies can lead to fragmented assemblies whenever a complex repeat or poorly amplified region is encountered. As a result, GC-rich and GC-poor regions, which tend to be poorly amplified, are particularly susceptible to poor quality sequencing. Resolving fragmented assemblies requires additional costly bench work and further sequencing. By also including the longer reads of SMRT sequencing runs, the read set will span many more repeats and missing bases, thereby closing many of the gaps automatically and simplifying, or even eliminating, the finishing time. It is becoming routine for bacterial genomes to be completely assembled using this approach, and we expect this practice will translate to larger genomes in the near future. A complete genome is far more useful than the poor quality draft sequences that litter GenBank because it provides a complete blueprint for the organism; the genes encoded therein represent the full biological potential of that organism. With only draft assemblies available, one is always left with the nagging feeling that some crucial gene is missing - perhaps the one in which you are most interested. The long read lengths also have more power to reveal complex structural variations present in DNA samples, such as pinpointing precisely where copy number variations have occurred relative to the reference sequence. They are also extremely powerful for resolving complex RNA splicing patterns from cDNA libraries, since a single long read may contain the entire transcript end-to-end, thus eliminating the need to infer the isoforms.

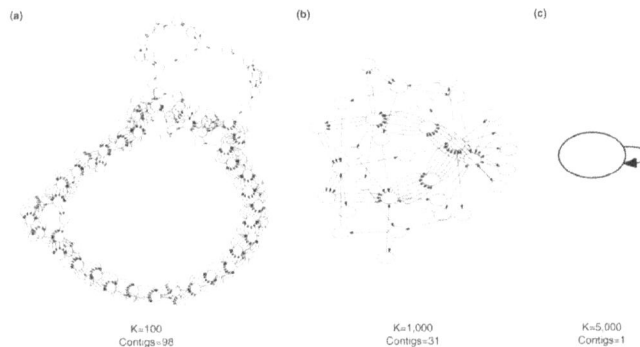

(a) K=100 Contigs=98   (b) K=1,000 Contigs=31   (c) K=5,000 Contigs=1

Idealized assembly graphs of the 5.2 megabase-pair *B. anthracis* Ames Ancestor main chromosome using (a) 100 bp, (b) 1,000 bp and (c) 5,000 bp reads. The graphs encode the compressed de Bruijn graph derived from infinite coverage error-free reads, effectively representing the repeats in the genome and the upper bound of what could be achieved in a real assembly. Increasing the read length decreases the number of contigs because the longer reads will span more of the repeats. Note the assembly with 5,000 bp reads has a self-edge because the chromosome is circular.

Second, consider DNA methyltransferases. These can exist as solitary entities or as parts of restriction-modification systems. In both cases, they methylate relatively short sequence motifs that can easily be recognized from SMRT sequencing data because of the change in DNA polymerase kinetics, as it moves along the template molecule, that result from the presence of epigenetic modifications. The altered kinetics cause a change in the timing of when the fluorescent colors are observed, thus enabling direct detection of epigenetic modifications, which can ordinarily only be inferred, and bypassing the usual necessity of enrichment or chemical conversion. Often, thanks to bioinformatics, the gene responsible for any given modification can be matched to the sequence motif in which the modification lies. When it cannot, then simply cloning the gene into a plasmid, which is subsequently grown in a non-modifying host and re-sequenced, can provide the match . Moreover, SMRT sequencing has also been able to identify RNA base modifications through the same approach as DNA base modifications, but using an RNA transcriptase in place of the DNA polymerase. In fact, SMRT sequencing represents an important step toward uncovering the biology that happens between DNA and proteins, including not only the study of mRNA sequences but also the regulation of translation. Thus, functional information emerges directly from the SMRT sequencing approach.

Third, we must consider the persistent rumor that SMRT sequencing is much less accurate than other next-generation sequencing platforms, which has now been demonstrated to be untrue in several ways. First, a direct comparison of several approaches to determining genetic polymorphisms has shown that SMRT sequencing has comparable performance to other sequencing technologies. Second, the accuracy of assembling a complete genome using SMRT sequencing in combination with other technologies has proved to be as reliable and accurate as more traditional approaches. Moreover Chin *l*. showed that an assembly using only long SMRT sequencing reads achieves comparable or even higher performance than other platforms (99.999% accuracy in three organisms with known reference sequences), including 11 corrections to the Sanger reference of these genomes. Koren showed that most microbial genomes could be assembled into a single contig per chromosome with this approach; it is by far the least expensive option for doing so.

Maxam and Gilbert Method

In 1977, Maxam and Gilbert described a sequencing method based on chemical degradation at specific locations of the DNA molecule. The end labeled DNA fragments are subjected to random cleavage at adenine, cytosine, guanine or thymine positions using specific chemical agents and the products of these fours reactions are separated using polyacrylamide gel electrophoresis (PAGE). As in Sanger method, the sequence can be easily read from four parallel lanes in the sequencing gel.

Double stranded or single stranded DNA from chromosomal DNA can be used as template. Originally, end labeling was done with P phosphate or with a nucleotide linked to P and enzymatically incorporated into the end fragment. The read length is up to 500bp. The chemical reactions in the technique are slow and involved hazardous chemicals that require special handling in the DNA cleavage reaction.

As in Sanger's method, additional cautions in Maxam and Gilbert method include purification and separation of DNA fragments and higher analysis time. Therefore, this technology is not suitable for high throughput large-scale investigation.

Pal Nyren's Method

In 1996, Pal Nyren's group reported that natural nucleotide can be used to obtain efficient incorporation during a sequencing-by-synthesis protocol. The detection was based on the pyrophosphate (inorganic biphosphate) released during the DNA polymerase reaction, the quantitative conversion of pyrophosphate to ATP by sulfurylase and the subsequent production of visible light by firefly luciferase.

The first major improvement was inclusion of dATPaS in place of dATP in the polymerization reaction, which enabled the pyrosequencing reaction to be performed in homogeneous phase in real time.

The non-specific signals were attributed to the fact that dATP is a substrate for luciferase. Conversely, dATPaS was found to be inert for luciferase, yet could be incorporated efficiently by all DNA polymerases tested. The second improvement was the introduction apyrase to the reaction to make a four-enzyme system. Apyrase allows nucleotides to be added sequentially without any intermediate washing step.

Pyrosequencing nonelectrophoretic real-time DNA sequencing method is based on sequencing by synthesis based on the pyrophosphate (inorganic biphosphate) released during the DNA polymerase reaction.

In a cascade of enzymatic reaction, visible light is generated that is proportional to the number of incorporated nucleotides. The cascade starts with a nucleic acid polymerization reaction in which inorganic bip-hosphate (PPi) is released as a result of nucleotide incorporation by polymerase.

The released PPi is subsequently converted to ATP by ATP sulfurylase, which provides the energy to luciferase to oxidize luciferin and generate light. The light so generated is captured by a CCD camera and recorded in the form of peaks known as pyrogram (compared with electropherograms in Sanger's method). Because the added nucleotide is known the sequence of template can be determined.

Standard pyrosequencing uses the Klenow fragment of E. coli DNA pol I, which is relatively slow polymerase. The ATP sulfurylase used in pyrosequencing is a recombinant version from the yeast and the luciferase is from the American firefly. The overall reaction from polymerization to light detection takes place within three to four seconds at real time.

One pmol of DNA in a pyrosequencing reaction yields 6×10^{11} ATP molecules which in turn, generate more than 6×10^{9} photons at a wavelength of 560 nm. This amount of light is easily detected by a photodiode, photomultiplier tube or a CCD camera. Pyrosequencing technology has been further improved into array-based massively parallel microfluidic sequencing platform.

Automatic DNA Sequencer

A variant of the above dideoxy-method was developed, which allowed the production of automatic sequencers. In this new approach, different fluorescent dyes are tagged either to the oligonucleotide primer (dye primers) in each of the four reaction tubes (blue for A, red for C, etc), or to each of the four ddNTPs (dye terminators) used in a single reaction tube: when four tubes are used, they are pooled.

After the PCR reaction is over, the reaction mixture is subjected to separation of synthesized fragments through electrophoresis. Depending upon the electrophoretic system used, whether slab gel electrophoresis or capillary electrophoresis, following two types of automatic sequencing systems have been designed.

Slab Gel Sequencing Systems:

These systems make use of ultrathin (75 μm) slab gels and involve running of atleast 96 lanes per gel. In these systems, automation in sample loading of sequencing gels has also been achieved, by using a plexiglass block having wells that are same distance apart as the comb teeth cut in a porous membrane that is used as a comb for drawing samples by capillary action.

Each well in plexiglass block is filed with a sample (PCR dideoxy-reaction mixture), so that when the porous membrane comb is lowered onto the sample wells in the pexiglass the samples are drawn up automatically into the comb teeth by capillary action.

Using this approach of employing porous combs, automated loading of up to 192, 384 or 480 samples per gel has been achieved. The porous comb with the samples is placed between the glass plates of the gel apparatus above the flat surface of the polymerized gel and the samples are driven from the comb into the gel by electrophoresis.

Capillary Gel Electrophoresis:

In these systems, slab gel electrophoresis is replaced by capillary gel electrophoresis to analyse DNA samples. In these systems, instead of scanning DNA as it migrates through 96 lanes each in a series of 96 capillary tubes, DNA fragments pass are scanned.

In the original models of the above old slab gel machines, gels must be poured and reagents frequently reloaded, interrupting the sequencing.

In capillary gel sequencing systems, on the other hand, the robot moves the DNA samples and reagents through the tubes continuously, requiring attention only once a day. The system produces a steady flow of data, each signal representing one of the four DNA bases (adenine, cytosine, guanine and thymine).

RNA Sequencing

RNA-Seq is a recently developed approach to transcriptome profiling that uses deep-sequencing technologies. Studies using this method have already altered our view of the extent and complexity of eukaryotic transcriptomes. RNA-Seq also provides a far more precise measurement of levels of transcripts and their isoforms than other methods.

The transcriptome is the complete set of transcripts in a cell, and their quantity, for a specific developmental stage or physiological condition. Understanding the transcriptome is essential for interpreting the functional elements of the genome and revealing the molecular constituents of cells and tissues, and also for understanding development and disease. The key aims of transcriptomics are: to catalogue all species of transcript, including mRNAs, non-coding RNAs and small RNAs; to determine the transcriptional structure of genes, in terms of their start sites, 5′ and 3′ ends, splicing patterns and other post-transcriptional modifications; and to quantify the changing expression levels of each transcript during development and under different conditions.

Various technologies have been developed to deduce and quantify the transcriptome, including hybridization-or sequence-based approaches. Hybridization-based approaches typically involve incubating fluorescently labelled cDNA with custom-made microarrays or commercial high-density oligo microarrays. Specialized microarrays have also been designed; for example, arrays with probes spanning exon junctions can be used to detect and quantify distinct spliced isoforms[1]. Genomic tiling microarrays that represent the genome at high density have been constructed and allow the mapping of transcribed regions to a very high resolution, from several base pairs to ~100 bp. Hybridization-based approaches are high throughput and relatively inexpensive, except for high-resolution tiling arrays that interrogate large genomes. However, these methods have several limitations, which include: reliance upon existing knowledge about genome sequence; high background levels owing to cross-hybridization; and a limited dynamic range of detection owing to both background and saturation of signals. Moreover, comparing expression levels across different experiments is often difficult and can require complicated normalization methods.

In contrast to microarray methods, sequence-based approaches directly determine the cDNA sequence. Initially, Sanger sequencing of cDNA or EST libraries was used, but this approach is relatively low throughput, expensive and generally not quantitative. Tag-based methods were developed to overcome these limitations, including serial analysis of gene expression (SAGE), cap analysis of gene expression (CAGE) and massively parallel signature sequencing (MPSS). These tag-based sequencing approaches are high throughput and can provide precise, 'digital' gene expression levels. However, most are based on expensive Sanger sequencing technology, and a significant portion of the short tags cannot be uniquely mapped to the reference genome. Moreover, only a portion of the transcript is analysed and isoforms are generally indistinguishable from each other. These disadvantages limit the use of traditional sequencing technology in annotating the structure of transcriptomes.

Recently, the development of novel high-throughput DNA sequencing methods has provided a new method for both mapping and quantifying transcriptomes. This method, termed RNA-Seq (RNA sequencing), has clear advantages over existing approaches and is expected to revolutionize the manner in which eukaryotic transcriptomes are analysed. It has already been applied to *Saccharomyces cerevisiae, Schizosaccharomyces pombe, Arabidopsis thaliana*, mouse and human cells.

RNA-Seq Technology and Benefits

RNA-Seq uses recently developed deep-sequencing technologies. In general, a population of RNA (total or fractionated, such as poly (A)+) is converted to a library of cDNA fragments with adaptors attached to one or both ends. Each molecule, with or without amplification, is then sequenced in a high-throughput manner to obtain short sequences from one end (single-end sequencing) or both ends (pair-end sequencing).The reads are typically 30–400 bp, depending on the DNA-sequencing technology used. In principle, any high-throughput sequencing technology can be used for RNA-Seq, and the Illumina IG, Applied Biosystems SOLiD and Roche 454 Life Science systems have already been applied for this purpose. The Helicos Biosciences tSMS system has not yet been used for published RNA-Seq studies, but is also appropriate and has the added advantage of avoiding amplification of target cDNA. Following sequencing, the resulting reads are either aligned to a reference genome or reference transcripts, or assembled *de novo* without the genomic sequence to produce a genome-scale transcription map that consists of both the transcriptional structure and/or level of expression for each gene.

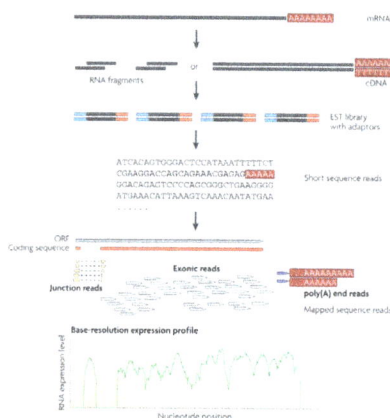

A typical RNA-Seq experiment

Briefly, long RNAs are first converted into a library of cDNA fragments through either RNA fragmentation or DNA fragmentation. Sequencing adaptors (blue) are subsequently added to each cDNA fragment and a short sequence is obtained from each cDNA using high-throughput sequencing technology. The resulting sequence reads are aligned with the reference genome or transcriptome, and classified as three types: exonic reads, junction reads and poly(A) end-reads. These three types are used to generate a base-resolution expression profile for each gene, as illustrated at the bottom; a yeast ORF with one intron is shown.

Although RNA-Seq is still a technology under active development, it offers several key advantages over existing technologies. First, unlike hybridization-based approaches, RNA-Seq is not limited to detecting transcripts that correspond to existing genomic sequence. For example, 454-based RNA-Seq has been used to sequence the transcriptome of the Glanville fritillary butterfly. This makes RNA-Seq particularly attractive for non-model organisms with genomic sequences that are yet to be determined. RNA-Seq can reveal the precise location of transcription boundaries, to a single-base resolution. Furthermore, 30-bp short reads from RNA-Seq give information about how two exons are connected, whereas longer reads or pair-end short reads should reveal connectivity between multiple exons. These factors make RNA-Seq useful for studying complex transcriptomes. In addition, RNA-Seq can also reveal sequence variations (for example, SNPs) in the transcribed regions.

Advantages of RNA-Seq compared with other transcriptomics methods

Technology	Tiling microarray	cDNA or EST sequencing	RNA-seq
Technology specifications			
Principle	Hybridization	Sanger sequencing	High-throughput sequencing
Resolution	From several to 100 bp	Single base	Single base
Throughput	High	Low	High
Reliance on genomic sequence	Yes	No	In some cases
Background noise	High	Low	Low
Application			

Technology	Tiling microarray	cDNA or EST sequencing	RNA-seq
Simultaneously map transcribed regions and gene expression	Yes	Limited for gene expression	Yes
Dynamic range to quantify gene expression level	Up to a few-hundred-fold	Not practical	>8,000-fold
Ability to distinguish different isoforms	Limited	Yes	Yes
Ability to distinguish allelic expression	Limited	Yes	Yes
Practical issues			
Required amount of RNA	High	High	Low
cost for mapping transcriptomes of large genomes	High	High	Relatively low

A second advantage of RNA-Seq relative to DNA microarrays is that RNA-Seq has very low, if any, background signal because DNA sequences can been unambiguously mapped to unique regions of the genome. RNA-Seq does not have an upper limit for quantification, which correlates with the number of sequences obtained. Consequently, it has a large dynamic range of expression levels over which transcripts can be detected: a greater than 9,000-fold range was estimated in a study that analyzed 16 million mapped reads in *Saccharomyces cerevisiae*, and a range spanning five orders of magnitude was estimated for 40 million mouse sequence reads. By contrast, DNA microarrays lack sensitivity for genes expressed either at low or very high levels and therefore have a much smaller dynamic range (one-hundredfold to a few-hundredfold). RNA-Seq has also been shown to be highly accurate for quantifying expression levels, as determined using quantitative PCR (qPCR) and spike-in RNA controls of known concentration. The results of RNA-Seq also show high levels of reproducibility, for both technical and biological replicates. Finally, because there are no cloning steps, and with the Helicos technology there is no amplification step, RNA-Seq requires less RNA sample.

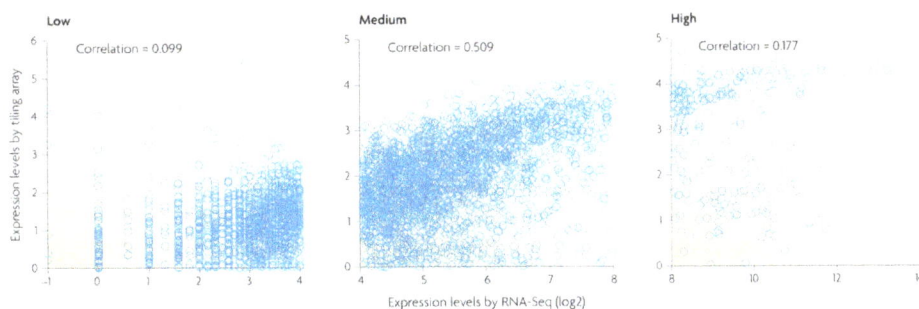

Quantifying Expression levels: RNA-Seq and Microarray Compared

Expression levels are shown, as measured by RNA-Seq and tiling arrays, for *Saccharomyces cerevisiae* cells grown in nutrient-rich media. The two methods agree fairly well for genes with medium levels of expression (middle), but correlation is very low for genes with either low or high expression levels. The tiling array data used in this figure is taken from REF. and the RNA-Seq data is taken from REF.

Taking all of these advantages into account, RNA-Seq is the first sequencing-based method that allows the entire transcriptome to be surveyed in a very high-throughput and quantitative manner. This method offers both single-base resolution for annotation and 'digital' gene expression levels at the genome scale, often at a much lower cost than either tiling arrays or large-scale Sanger EST sequencing.

Challenges for RNA-Seq

Library Construction

The ideal method for transcriptomics should be able to directly identify and quantify all RNAs, small or large. Although there are only a few steps in RNA-Seq, it does involve several manipulation stages during the production of cDNA libraries, which can complicate its use in profiling all types of transcript.

Unlike small RNAs (microRNAs (miRNAs), Piwi-interacting RNAs (piRNAs), short interfering RNAs (siRNAs) and many others), which can be directly sequenced after adaptor ligation, larger RNA molecules must be fragmented into smaller pieces (200–500 bp) to be compatible with most deep-sequencing technologies. Common fragmentation methods include RNA fragmentation (RNA hydrolysis or nebulization) and cDNA fragmentation (DNase I treatment or sonication). Each of these methods creates a different bias in the outcome. For example, RNA fragmentation has little bias over the transcript body, but is depleted for transcript ends compared with other methods. Conversely, cDNA fragmentation is usually strongly biased towards the identification of sequences from the 3′ ends of transcripts, and thereby provides valuable information about the precise identity of these ends.

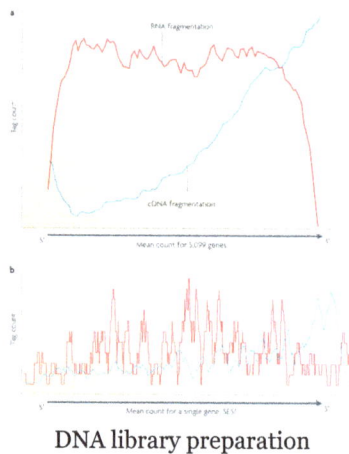

DNA library preparation

RNA fragmentation and DNA fragmentation compared a Fragmentation of oligo-dT primed cDNA (blue line) is more biased towards the 3′ end of the transcript. RNA fragmentation (red line) provides more even coverage along the gene body, but is relatively depleted for both the 5′ and 3′ ends. Note that the ratio between the maximum and minimum expression level (or the dynamic range) for microarrays is 44, for RNA-Seq it is 9,560. The tag count is the average sequencing coverage for 5,000 yeast ORFs18. b | A specific yeast gene, SES1 (seryl-tRNA synthetase), is shown.

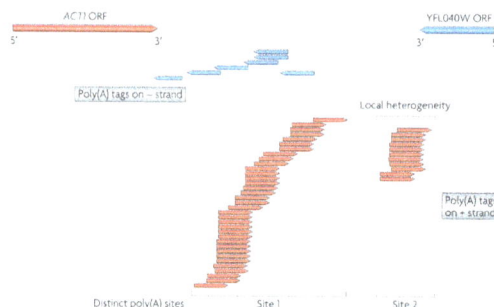

Poly (A) tags from RNA-Seq

A region containing two overlapping transcripts (*ACT1*, from the actin gene, and YFL040W, an uncharacterized ORF) from the *Saccharomyces cerevisiae* genome is shown. Arrows point to transcription direction. The poly (A) tags from RNA-Seq experiments are shown below these transcripts, with arrows indicating transcription direction. The precise location of each locus identified by poly (A) tags reveals the heterogeneity in poly (A) sites, for example, *ACT1* has two big clusters, both with a few bases of local heterogeneity. The transcription direction revealed by poly(A) tags also helps to resolve 3'-end overlapping transcribed regions.

Some manipulations during library construction also complicate the analysis of RNA-Seq results. For example, many shorts reads that are identical to each other can be obtained from cDNA libraries that have been amplified. These could be a genuine reflection of abundant RNA species, or they could be PCR artefacts. One way to discriminate between these possibilities is to determine whether the same sequences are observed in different biological replicates.

Another key consideration concerning library construction is whether or not to prepare strand-specific libraries, as has been done in two studies. These libraries have the advantage of yielding information about the orientation of transcripts, which is valuable for transcriptome annotation, especially for regions with overlapping transcription from opposite directions; however, strand-specific libraries are currently laborious to produce because they require many steps or direct RNA–RNA ligation, which is inefficient. Moreover, it is essential to ensure that the antisense transcripts are not artefacts of reverse transcription. Because of these complications, most studies thus far have analyzed cDNAs without strand information.

Bioinformatic Challenges

Like other high-throughput sequencing technologies, RNA-Seq faces several informatics challenges, including the development of efficient methods to store, retrieve and process large amounts of data, which must be overcome to reduce errors in image analysis and base-calling and remove low-quality reads.

Once high-quality reads have been obtained, the first task of data analysis is to map the short reads from RNA-Seq to the reference genome, or to assemble them into contigs before aligning them to the genomic sequence to reveal transcription structure. There are several programs for mapping reads to the genome, including ELAND, SOAP, MAQ and RMAP (information about these can be found at the Illumina forum and at SEQanswers). However, short transcriptomic reads also contain reads that span exon junctions or that contain poly(A) ends — these cannot be analyzed in the same way. For genomes in which splicing is rare (for example, *S. cerevisiae*) special attention only needs to be given to poly (A) tails and to a small number of exon–exon junctions. Poly (A) tails can be identified simply by the presence of multiple As or Ts at the end of some reads. Exon–exon junctions can be identified by the presence of a specific sequence context (the GT–AG dinucleotides that flank splice sites) and confirmed by the low expression of intronic sequences, which are removed during splicing. Transcriptome maps have been generated in this manner for *S. cerevisiae*. For complex transcriptomes it is more difficult to map reads that span splice junctions, owing to the presence of extensive alternative splicing and *trans*-splicing. One partial solution is to compile a junction library that contains all the known and predicted junction sequences and map reads to this library. A challenge for the future is to develop computationally simple methods to identify novel splicing events that take place between two distant sequences or between exons from two different genes.

For large transcriptomes, alignment is also complicated by the fact that a significant portion of sequence reads match multiple locations in the genome. One solution is to assign these multi-matched reads by proportionally assigning them based on the number of reads mapped to their neighbouring unique sequences. This method has been successful for low-copy repetitive sequences. Short reads that have high copy numbers (>100) and long stretches of repetitive regions present a greater challenge. Obtaining longer sequence reads, for example using 454 technology, should help alleviate the multi-matching problem. Alternatively, a paired-end sequencing strategy, in which short sequences are determined from both ends of a DNA fragment, extends the mapped fragment length to 200–500 bp and is expected to be useful in the future. Sequencing errors and polymorphisms can present mapping problems for all genomes, not just for repetitive DNA. Generally, single base differences are not problematic, because most mapping algorithms accommodate one or two base differences. However, resolving larger differences will require better reference genome annotation for polymorphisms and deeper sequencing coverage.

Coverage Versus Cost

Another important issue is sequence coverage, or the percentage of transcripts surveyed, which has implications for cost. Greater coverage requires more sequencing depth. To detect a rare transcript or variant, considerable depth is needed. In simple transcriptomes, such as yeast for which there is no evidence of alternative splicing, 30 million 35-nucleotide reads from poly(A) mRNA libraries are sufficient to observe transcription from most (>90%) genes for cells grown under a single condition (that is, in nutrient-rich medium). This depth is probably more than sufficient for most purposes, as the number of expressed genes detected by RNA-Seq reaches 80% coverage at 4 million uniquely mapped reads, after which doubling the depth merely increases the coverage by 10%. The remaining genes are presumably either not expressed under this condition (for example, sporulation genes) or do not have poly(A) tails. Analyzing many different conditions can further increase the coverage; in *S. pombe* 122 million reads from six different growth conditions detected transcription from >99% of annotated genes.

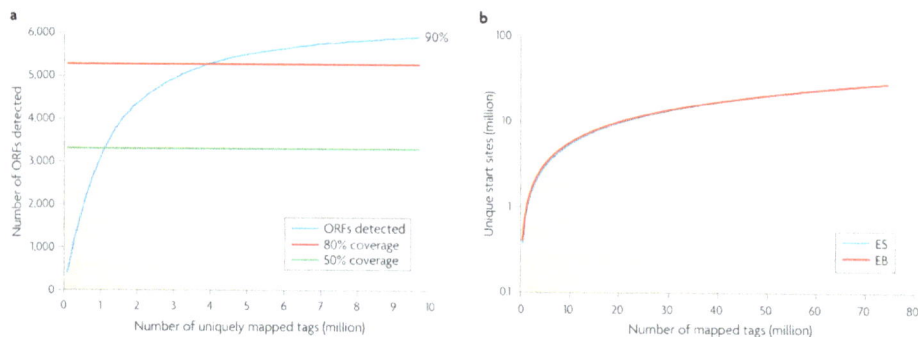

Coverage versus depth

a | 80% of yeast genes were detected at 4 million uniquely mapped RNA-Seq reads, and coverage reaches a plateau afterwards despite the increasing sequencing depth. Expressed genes are defined as having at least four independent reads from a 50-bp window at the 3′ end. Data is taken from REF. b | The number of unique start sites detected starts to reach a plateau when the depth of sequencing reaches 80 million in two mouse transcriptomes. ES, embryonic stem cells; EB, embryonic body.

In general, the larger the genome, the more complex the transcriptome, the more sequencing depth is required for adequate coverage. Unlike genome-sequencing coverage, it is less straightforward to calculate the coverage of the transcriptome; this is because the true number and level of different transcript isoforms is not usually known and because transcription activity varies greatly across the genome. One study used the number of unique transcription start sites as a measure of coverage in mouse embryonic cells, and demonstrated that at 80 million reads, the number of start sites reached a plateau. However, this approach does not address transcriptome complexity in alternative splicing and transcription termination sites; presumably further sequencing can reveal additional variants.

New Transcriptomic Insights

Despite the challenges described above, the advantages of RNA-Seq have enabled us to generate an unprecedented global view of the transcriptome and its organization for a number of species and cell types. Before the advent of RNA-Seq, it was known that a much greater than expected fraction of the yeast, *Drosophila melanogaster* and human genomes are transcribed, and for yeast and humans a number of distinct isoforms have been found for many genes. However, the starts and ends of most transcripts and exons had not been precisely resolved and the extent of spliced heterogeneity remained poorly understood. RNA-Seq, with its high resolution and sensitivity has revealed many novel transcribed regions and splicing isoforms of known genes, and has mapped 5′ and 3′ boundaries for many genes.

Mapping Gene and Exon Boundaries

The single-base resolution of RNA-Seq has the potential to revise many aspects of the existing gene annotation, including gene boundaries and introns for known genes as well as the identification of novel transcribed regions. 5′ and 3′ boundaries can be mapped to within 10–50 bases by a precipitous drop in signal. 3′ boundaries can be precisely mapped by searching for poly (A) tags, and introns can be mapped by searching for tags that span GT–AG splicing consensus sites. Using these methods the 5′ and 3′ boundaries of 80% and 85% of all annotated genes, respectively, were mapped in *S. cerevisiae*. Similarly, in *S. pombe* many boundaries were defined by RNA-Seq data in combination with tiling array data.

These two studies led to the discovery of many 5′ and 3′ UTRs that had not been analyzed previously. In *S. cerevisiae*, extensive 3′-end heterogeneity was discovered at two levels: first, local heterogeneity exists in which a cluster of sites are involved, typically within a 10 bp window; second, there are distinct regions of poly(A) addition for 540 genes. It is plausible that these different 3′ ends confer distinct properties to the different mRNA isoforms, such as mRNA localization or degradation signals, which in turn might be responsible for unique biological functions. In addition to 3′ heterogeneity, the list of upstream ORFs within the 5′ UTRs of mRNAs (uORFs) was also greatly expanded from 17 to 340 (6% of yeast genes); uORFs regulate mRNA translation or stability, so these sequences might make a previously underappreciated contribution to the regulatory sophistication of eukaryotic genomes. Interestingly, many mRNAs with uORFs are transcription factors, suggesting that these regulators are themselves heavily regulated.

The mapping of transcript boundaries revealed several novel features of eukaryotic gene organization. Many yeast genes were found to overlap at their 3′ ends. Using relaxed criteria similar to those employed in a recent study we found that 808 pairs, approximately 25% of all yeast ORFs, overlap at their 3′ ends. Likewise, antisense expression is enriched in the 3′ exons of mouse transcripts. These features might confer interesting regulatory properties on the affected genes. For multicellular organisms, antisense transcription could modulate gene expression through the production of siRNAs or through dsRNA editing. For yeast, which seems to lack siRNA and dsRNA-editing functions, transcription from one gene might interfere with that from an overlapping gene, or coordinate gene expression through other mechanisms.

Extensive Transcript Complexity

RNA-Seq can be used to quantitatively examine splicing diversity by searching for reads that span known splice junctions as well as potential new ones. In humans, 31,618 known splicing events were confirmed (11% of all known splicing events) and 379 novel splicing events were discovered. Another study of human cells found 94,241 junctions, among which 4,096 were novel, and further demonstrated that the prevalent form of alternative splicing is exon skipping. In mice, extensive alternative splicing was observed for 3,462 genes. In addition, 42 splicing events that join exons from multiple mouse genes were detected.

Novel Transcription

Previous studies using transposon tagging and tiling microarrays have suggested that in the genomes of yeast, *D. melanogaster* and humans, there are many novel transcribed regions represented in poly(A)+ RNA. However, the accuracy of the tiling array results is uncertain owing to concerns about cross-hybridization. RNA-Seq, which does not suffer from problems with background noise, has confirmed that at least 75% and perhaps greater than 90% of the *S. cerevisiae* and *S. pombe* genomes are expressed. In addition, results from RNA-Seq suggest the existence of a large number of novel transcribed regions in every genome surveyed, including the *A. thaliana*, mouse, human, *S. cerevisiae* and *S. pombe* genomes. 487 and 453 novel transcripts have been discovered in *S. cerevisiae* and *S. pombe*, respectively; for *S. cerevisiae* half of these were not identified using microarrays. Many of these novel transcribed regions in yeast do not seem to encode any protein, and their functions remain to be determined. The current sequencing depth is not sufficient to define the boundaries of novel transcript units in mammals; however, 30–40% of reads map to unannotated regions. These novel transcribed regions, combined with many undiscovered novel splicing variants, suggest that there is considerably more transcript complexity than previously appreciated.

Defining Transcription Level

As RNA-Seq is quantitative, it can be used to determine RNA expression levels more accurately than microarrays. In principle, it is possible to determine the absolute quantity of every molecule in a cell population, and directly compare results between experiments. Several methods have been used for quantification. For RNA fragmentation followed by cDNA synthesis, which gives more uniform coverage of each exon, gene expression levels can be deduced from the total number of reads that fall into the exons of a gene, normalized by the length of exons that can be uniquely mapped; for 3′-biased methods, read counts from a window near the 3′ end are used. Gene expression levels determined by these methods closely correlate with qPCR and RNA spike-in controls.

One particularly powerful advantage of RNA-Seq is that it can capture transcriptome dynamics across different tissues or conditions without sophisticated normalization of data sets". RNA-Seq has been used to accurately monitor gene expression during yeast vegetative growth, yeast meiosis and mouse embryonic stem-cell differentiation, to track gene expression changes during development, and to provide a 'digital measurement' of gene expression difference between different tissues. Because of these advantages, RNA-Seq will undoubtedly be valuable for understanding transcriptomic dynamics during development and normal physiological changes, and in the analysis of biomedical samples, where it will allow robust comparison between diseased and normal tissues, as well as the subclassification of disease states.

Future Directions

Although RNA-Seq is still in the early stages of use, it has clear advantages over previously developed transcriptomic methods. The next big challenge for RNA-Seq is to target more complex transcriptomes to identify and track the expression changes of rare RNA isoforms from all genes. Technologies that will advance achievement of this goal are pair-end sequencing, strand-specific sequencing and the use of longer reads to increase coverage and depth. As the cost of sequencing continues to fall, RNA-Seq is expected to replace microarrays for many applications that involve determining the structure and dynamics of the transcriptome.

References

- Dear PH (2003). "One by one: Single molecule tools for genomics". Briefings in Functional Genomics and Proteomics. 1 (4): 397–416. doi:10.1093/bfgp/1.4.397. PMID 15239886

- Campbell NA and Reece JB. (2002) Biology (6th ed.). San Francisco: Benjamin Cummings. ISBN 0-8053-6624-5

- Dna-sequencing-7-methods-used-for-dna-sequencing-11840: biologydiscussion.com, Retrieved 21 May 2018

- Oshima Y; Sawada, H.; Hosokawa, F.; Okunishi, E.; Kaneyama, T.; Kondo, Y.; Niitaka, S.; Takagi, H.; Tanishiro, Y.; Takayanagi, K.; et al. (2010). "Direct imaging of lithium atoms in LiV2O4 by spherical aberration-corrected electron microscopy". Journal of Electron Microscopy. 59 (6): 457–61. doi:10.1093/jmicro/dfq017. PMID 20406731

- Nanoball-sequencing: dmohankumar.wordpress.com, Retrieved 26 June 2018

- Xu, M; Fujita, Daisuke; Hanagata, Nobutaka; et al. (2009). "Perspectives and Challenges of Emerging Single-Molecule DNA Sequencing Technologies". Small. 5 (23): 2638–49. doi:10.1002/smll.200900976. PMID 19904762

- Dna-sequencing-7-methods-used-for-dna-sequencing-11840: biologydiscussion.com, Retrieved 31 March 2018

- Tucker T; Marra, Marco; Friedman, Jan M.; et al. (2009). "Massively Parallel Sequencing: The Next Big Thing in Genetic Medicine". The American Journal of Human Genetics. 85 (2): 142–54. doi:10.1016/j.ajhg.2009.06.022. PMC 2725244 . PMID 19679224

## 4

Bioinformatics Algorithms

Bioinformatics involves the development of algorithms, databases, and statistical and computational techniques for the solution of problems in the analysis of biological data. The aim of this chapter is to explore the varied algorithms used in bioinformatics, such as Baum–Welch algorithm, Needleman–Wunsch algorithm, String searching algorithm, Hirschberg's algorithm, Smith–Waterman algorithm, etc. which are crucial for a complete understanding of bioinformatics.

Baum–Welch Algorithm

A cornerstone of stochastic grammar-based profiling of DNA, protein and RNA sequences is the fast and accurate estimation of the probability parameters of the model, given a 'training set'. This is achieved for hidden Markov models (HMMs) using the Baum–Welch algorithm.

Let us consider discrete (categorical) HMMs of length T (each observation sequence is T observations long). Let the space of observations be $X = \{1, 2, ..., N\}$, and let the space of underlying states be $Z = \{1, 2, ..., M\}$. An HMM $\theta = (\pi, A, B)$ is parameterized by the initial state matrix π, the state transition matrix A, and the emission matrix B; $\pi_i = P(z_1 = i)$, $A_{ij} = P(z_{t+1} = j \mid z_t = 1)$, and $B_i(j) = P(x_t = j \mid zt = i)$.

We study the problem of learning the parameterization of θ from a dataset of D observations. Let $\chi = \left(X^{(1)}, ..., X^{(D)}\right)$, where each $X^{(i)} = \left(x_1^{(i)}, x_2^{(i)}, ..., x_T^{(i)}\right)$ We assume each observation is drawn iid. The learning problem is nontrivial because we are not given the latent variable $Z^{(i)}$ for each $X^{(i)}$, otherwise we could directly compute $\theta^* = argmax_\theta\, P(\mathcal{X}, \mathcal{Z}; \theta)$ Without $\mathcal{Z}$, the naive solution would be to directly compute $\theta^* = argmax_\theta\, \sum_{z \in \mathcal{Z}} P(\mathcal{X}, z; \theta)$ This is not tractable, since there are DT^M different values of z to try.

Baum-Welch is an iterative procedure for estimating θ^* from only $\mathcal{X}$. It works by maximizing a proxy to the log-likelihood, and updating the current model to be closer to the optimal model. Each iteration of Baum-Welch is guaranteed to increase the log-likelihood of the data. But of course, convergence to the optimal solution is not guaranteed. Baum-Welch can be described simply as repeating the following steps until convergence:

1. Compute $Q(\theta, \theta^s) = \sum_{z \in \mathcal{Z}} \log \left[P(\mathcal{X}, z; \theta)\right] P(z \mid X; \theta^s).$

2. Set $\theta^{s+1} = \underset{\theta}{argmax}\, Q(\theta, \theta^s).$

Without justifying why this works, the rest of this document will focus on deriving the necessary update steps to run this algorithm. First, noting that $P(z, \mathcal{X}) = P(\mathcal{X})P(z \mid \mathcal{X})$, we can write

$$\operatorname*{argmax}_{\theta}\sum_{z\in Z}\log\left[P(\mathcal{X},z;\theta)\right]P(z\mid\mathcal{X};\theta^s)=\operatorname*{argmax}_{\theta}\sum_{z\in Z}\log\left[P(\mathcal{X},z;\theta)\right]P(z\mid\mathcal{X};\theta^s)=\operatorname*{argmax}_{\theta}\hat{Q}(\theta,\theta^s)$$

since $P(\mathcal{X})$ is not affected by choice of θ. Now $P(z,\mathcal{X};\theta)$ is easy to write down

$$P(z,\mathcal{X};\theta)=\prod_{d=1}^{D}\left(\pi_{z_1^{(d)}}B_{z_1^{(d)}}\left(x_1^{(d)}\right)\prod_{t=2}^{T}A_{z_{t-1}^{(d)}z_t^{(d)}}B_{z_t^{(d)}}\left(x_t^{(d)}\right)\right)$$

Taking the log gives us

$$\log P(z,\mathcal{X};\theta)=\sum_{d=1}^{D}\left[\log\pi_{z_1^{(d)}}+\sum_{t=2}^{T}\log A_{z_{t-1}^{(d)}z_t^{(d)}}+\sum_{t=1}^{T}\log B_{z_t^{(d)}}\left(x_t^{(d)}\right)\right]$$

Plugging this into $\hat{Q}(\theta,\theta^s)$, we get

$$\hat{Q}(\theta,\theta^s)=\sum_{z\in Z}\sum_{d=1}^{D}\log\pi_{z_1^{(d)}}P(z,\mathcal{X};\theta^s)+\sum_{z\in Z}\sum_{d=1}^{D}\sum_{t=2}^{T}\log A_{z_{t-1}^{(d)}z_t^{(d)}}P(z,\mathcal{X};\theta^s)+\sum_{z\in Z}\sum_{d=1}^{D}\sum_{t=1}^{T}\log B_{z_t^{(d)}}\left(x_t^{(d)}\right)P(z,\mathcal{X};\theta^s)$$

This is a nice form which we can optimize analytically with Lagrange multipliers. We need Lagrange multipliers because we have equality constraints which come from requiring that π, $A_i\cdot$ and $B_i(\cdot)$ form valid probability distributions. Let $\hat{L}(\theta,\theta^s)$ be the Lagrangian

$$\hat{L}(\theta,\theta^s)=\hat{Q}(\theta,\theta^s)-\lambda_\pi\left(\sum_{i=1}^{M}\pi_i-1\right)-\sum_{i=1}^{M}\lambda_{A_i}\left(\sum_{i=1}^{M}A_{ij}-1\right)-\sum_{i=1}^{M}\lambda_{B_i}\left(\sum_{i=1}^{N}B_i(j)-1\right)$$

First let us focus on the π_i 's

$$\frac{\partial\hat{L}(\theta,\theta^s)}{\partial\pi_i}=\frac{\partial}{\partial\pi_i}\left[\sum_{z\in Z}\sum_{d=1}^{D}\log\pi_{z_1^{(d)}}P(z,X;\theta^s)\right]-\lambda_\pi=0$$

$$=\frac{\partial}{\partial\pi_i}\left[\sum_{j=1}^{M}\sum_{d=1}^{D}\log\pi_j P\left(z_1^{(d)}=j,\mathcal{X};\theta^s\right)\right]-\lambda_\pi=0$$

$$=\sum_{d=1}^{D}\frac{P\left(z_1^{(d)}=i,\mathcal{X};\theta^s\right)}{\pi_i}-\lambda_\pi=0$$

$$\frac{\partial\hat{L}(\theta,\theta^s)}{\partial\lambda_\pi}=-\left(\sum_{j=1}^{M}\pi_i-1\right)=0$$

The second step is simply the result of marginalizing out, for each d, all $z_{z_t\neq1}^{(d)}$ and $z_t^{(d'\neq d)}$ for all t. Some algebra yields

$$\pi_i=\frac{\sum_{d=1}^{D}P\left(z_1^{(d)}=i,\mathcal{X};\theta^s\right)}{\sum_{j=1}^{M}\sum_{d=1}^{D}P\left(z_1^{(d)}=j,\mathcal{X};\theta^s\right)}=\frac{\sum_{d=1}^{D}P\left(z_1^{(d)}=i,\mathcal{X};\theta^s\right)}{\sum_{d=1}^{M}\sum_{j=1}^{D}P\left(z_1^{(d)}=j,\mathcal{X};\theta^s\right)}$$

$$= \frac{\sum\limits_{d=1}^{D} P\left(z_1^{(d)}=i, \mathbf{X} ; \theta^s\right)}{\sum\limits_{d=1}^{D} P\left(\mathbf{X} ; \theta^s\right)} = \frac{\sum\limits_{d=1}^{D} P\left(z_1^{(d)}=i, \mathbf{X} ; \theta^s\right)}{D\, P\left(\mathbf{X} ; \theta^s\right)}$$

$$= \frac{\sum\limits_{d=1}^{D} P\left(\mathbf{X} ; \theta^s\right) P\left(z_1^{(d)}=i \mid \mathbf{X} ; \theta^s\right)}{D\, P\left(\mathbf{X} ; \theta^s\right)} = \frac{1}{D} \sum\limits_{d=1}^{D} P\left(z_1^{(d)}=i \mid \mathbf{X} ; \theta^s\right)$$

$$= \frac{1}{D} \sum\limits_{d=1}^{D} P\left(z_1^{(d)}=i \mid X^{(d)} ; \theta^s\right)$$

We now follow a similar process for the A_{ij} 's.

$$\frac{\partial \hat{L}\left(\theta, \theta^s\right)}{\partial A_{ij}} = \frac{\partial}{\partial A_{ij}} \left[\sum_{z \in Z} \sum_{d=1}^{D} \sum_{t=2}^{T} \log A_{z_{t-1}^{(d)}\, z_t^{(d)}} P\left(z, \mathcal{X} ; \theta^s\right) \right] - \lambda_{A_i} = 0$$

$$= \frac{\partial}{\partial A_{ij}} \left[\sum_{j=1}^{M} \sum_{k=1}^{M} \sum_{d=1}^{D} \sum_{t=2}^{T} \log A_{jk} P\left(z_{t-1}^{(d)}=j, z_t^{(d)}=k, \mathcal{X} ; \theta^s\right) \right] - \lambda_{A_i} = 0$$

$$= \sum_{d=1}^{D} \sum_{t=2}^{T} \frac{P\left(z_{t-1}^{(d)}=i, z_t^{(d)}=j, \mathcal{X} ; \theta^s\right)}{A_{ij}} - \lambda_{A_i} = 0$$

$$\frac{\partial \hat{L}\left(\theta, \theta^s\right)}{\partial A_i} = -\left[\sum_{j=1}^{M} A_{ij} - 1 \right] = 0$$

This yields

$$A_{ij} = \frac{\sum\limits_{d=1}^{D} \sum\limits_{t=2}^{T} P\left(z_{t-1}^{(d)}=i, z_t^{(d)}=j, \mathcal{X} ; \theta^s\right)}{\sum\limits_{j=1}^{M} \sum\limits_{d=1}^{D} \sum\limits_{t=2}^{T} P\left(z_{t-1}^{(d)}=i, z_t^{(d)}=j, \mathcal{X} ; \theta^s\right)}$$

$$= \frac{\sum\limits_{d=1}^{D} \sum\limits_{t=2}^{T} P\left(z_{t-1}^{(d)}=i, z_t^{(d)}=j, \mathcal{X} ; \theta^s\right)}{\sum\limits_{d=1}^{D} \sum\limits_{t=2}^{T} P\left(z_{t-1}^{(d)}=i, \mathcal{X} ; \theta^s\right)}$$

$$= \frac{\sum\limits_{d=1}^{D} \sum\limits_{t=2}^{T} P\left(\mathcal{X} ; \theta^s\right) P\left(z_{t-1}^{(d)}=i, z_t^{(d)}=j \mid \mathcal{X} ; \theta^s\right)}{\sum\limits_{d=1}^{D} \sum\limits_{t=2}^{T} P\left(\mathcal{X} ; \theta^s\right) P\left(z_{t-1}^{(d)}=i \mid \mathcal{X} ; \theta^s\right)}$$

$$= \frac{\sum\limits_{d=1}^{D} \sum\limits_{t=2}^{T} P\left(z_{t-1}^{(d)}=i, z_t^{(d)}=j \mid \mathcal{X}^{(d)} ; \theta^s\right)}{\sum\limits_{d=1}^{D} \sum\limits_{t=2}^{T} P\left(z_{t-1}^{(d)}=i \mid \mathcal{X}^{(d)} ; \theta^s\right)}$$

The final thing is the $B_i(j)$'s, which are slightly trickier. Let $I(x)$ denote an indicator function which is 1 if x is true, 0 otherwise.

$$\frac{\partial \hat{L}(\theta, \theta^s)}{\partial B_i(j)} = \frac{\partial}{\partial B_i(j)} \left[\sum_{z \in Z} \sum_{d=1}^{D} \sum_{t=1}^{T} \log B_{z_t^{(d)}}\left(x_t^{(d)}\right) P(z, \mathcal{X} ; \theta^s) \right] - \lambda_{B_i} = 0$$

$$= \frac{\partial}{\partial B_i(j)} \left[\sum_{i=1}^{N} \sum_{d=1}^{D} \sum_{t=1}^{T} \log B_i\left(x_t^{(d)}\right) P\left(z_t^{(d)} = i, \mathcal{X} ; \theta^s\right) \right] - \lambda_{B_i} = 0$$

$$= \sum_{d=1}^{D} \sum_{t=1}^{T} \frac{P\left(z_t^{(d)} = i, \mathcal{X} ; \theta^s\right) I\left(x_t^{(d)} = j\right)}{B_i(j)} - \lambda_{B_i} = 0$$

$$\frac{\partial \hat{L}(\theta, \theta^s)}{\partial \lambda_{B_i}} = - \left[\sum_{j=1}^{N} B_i(j) - 1 \right] = 0$$

This should come as no surprise by now

$$B_i(j) = \frac{\sum_{d=1}^{D} \sum_{t=1}^{T} P\left(z_t^{(d)} = i, \mathcal{X} ; \theta^s\right) I\left(x_t^{(d)} = j\right)}{\sum_{i=1}^{N} \sum_{d=1}^{D} \sum_{t=1}^{T} P\left(z_t^d = i, \mathcal{X} ; \theta^s\right) I\left(x_t^{(d)} = j\right)}$$

$$= \frac{\sum_{d=1}^{D} \sum_{t=1}^{T} P\left(z_t^{(d)} = i, \mathcal{X} ; \theta^s\right) I\left(x_t^{(d)} = j\right)}{\sum_{d=1}^{D} \sum_{t=1}^{T} P\left(z_t^{(d)} = i, \mathcal{X}^{(d)} ; \theta^s\right)}$$

$$= \frac{\sum_{d=1}^{D} \sum_{t=1}^{T} P\left(z_t^{(d)} = i \mid X^{(d)} ; \theta^s\right) I\left(x_t^{(d)} = j\right)}{\sum_{d=1}^{D} \sum_{t=1}^{T} P\left(z_t^{(d)} = i \mid X^{(d)} ; \theta^s\right)}$$

To summarize, the update steps are

$$\pi_i^{(s+1)} = \frac{1}{D} \sum_{d=1}^{D} P\left(z_t^{(d)} = i \mid X^{(d)} ; \theta^s\right)$$

$$A_{ij}^{(s+1)} = \frac{\sum_{d=1}^{D} \sum_{t=2}^{T} P\left(z_{t-1}^{(d)} = i, z_t^{(d)} = j \mid X^{(d)} ; \theta^s\right)}{\sum_{d=1}^{D} \sum_{t=2}^{T} P\left(z_{t-1}^{(d)} = i, \mid X^{(d)} ; \theta^s\right)}$$

$$B_i^{(s+1)}(j) = \frac{\sum_{d=1}^{D} \sum_{t=1}^{T} P\left(z_t^{(d)} = i \mid X^{(d)} ; \theta^s\right) I\left(x_t^{(d)} = j\right)}{\sum_{d=1}^{D} \sum_{t=1}^{T} P\left(z_t^{(d)} = i \mid X^{(d)} ; \theta^s\right)}$$

Note that $P(z_t \mid X; \theta)$ and $P(z_{t-1}, z_t \mid X; \theta)$ are both quantities which can be computed efficiently for HMMs by the forward-backwards algorithm.

Needleman–Wunsch Algorithm

Dynamic programming algorithms find the best solution by breaking the original problem into smaller sub-problems and then solving. The Needleman-Wunsch algorithm is a dynamic programming algorithm for optimal sequence alignment. Basically, the concept behind the Needleman-Wunsch algorithm stems from the observation that any partial sub-path that tends at a point along the true optimal path must itself be the optimal path leading up to that point. Therefore the optimal path can be determined by incremental extension of the optimal sub-paths. In a Needleman-Wunsch alignment, the optimal path must stretch from beginning to end in both sequences.

Working of Needleman -Wunsch Algorithm

To study the algorithm, consider the two given sequences.

CGTGAATTCAT (sequence 1), GACTTAC (sequence 2)

The length (count of the nucleotides or amino acids) of the sequence 1 and sequence 2 are 11 and 7 respectively. The initial matrix is created with A+1 column's and B+1 row's (where A and B corresponds to length of the sequences). Extra row and column is given, so as to align with gap, at the starting of the matrix as shown in figure.

	-	C	G	T	G	A	A	T	T	C	A	T
-												
G												
A												
C												
T												
T												
A												
C												

Figure: Initial matrix

After creating the initial matrix, scoring schema has to be introduced which can be user defined with specific scores. The simple basic scoring schema can be assumed as, if two residues (nucleotide or amino acid) at i^{th} and j^{th} position are same, matching score is 1 $\left(S(i,j)=1\right)$ or if the two residues at i^{th} and j^{th} position are not same, mismatch score is assumed as -1 $\left(S(i,j)=-1\right)$. The gap score (w) or gap penalty is assumed as -1 .

Gap score is defined as penalty given to alignment, when we have insertion or deletion.

The dynamic programming matrix is defined with three different steps.

1. Initialization of the matrix with the scores possible.

2. Matrix filling with maximum scores.

3. Trace back the residues for appropriate alignment.

Initialization Step

This example assumes that there is gap penalty. First row and first column of the matrix can be initially filled with 0. If the gap score is assumed, the gap score can be added to the previous cell of the row or column.

	-	C	G	T	G	A	A	T	T	C	A	T
-	0	-1	-2	-3	-4	-5	-6	-7	-8	-9	-10	-11
G	-1											
A	-2											
C	-3											
T	-4											
T	-5											
A	-6											
C	-7											

Initialization of matrix

Matrix Fill Step

The second and crucial step of the algorithm is matrix filling starting from the upper left hand corner of the matrix. To find the maximum score of each cell, it is required to know the neighbouring scores (diagonal, left and right) of the current position. From the assumed values, add the match or mismatch (assumed) score to the diagonal value. Similarly add the gap score to the other neighbouring values. Thus, we can obtain three different values, from that take the maximum among them and fill the ith and j^{th} position with the score obtained.

In terms of matrix positions, it is important to know $[M_{(i-1,j-1)} + S_{(i,j)}, M_{(i,j-1)} + w, M_{(i-1,j)} + w]$

Overall the equation can be showed in the following manner

$$M_{i,j} = \text{Maximum } [M_{i-1,j-1} + S_{i,j}, M_{i,j-1} + w, M_{i-1,j} + w]$$

To score the matrix of the current position (the first position $M_{1,1}$) the above stated formulae can be used. The first residue (nucleotides or amino acids) in the 2 sequences are 'G' and 'C'. Since they are mismatching residues, the score would ($S_{i,j} = -1$) be -1.

$$M_{1,1} = Max \left[M_{0,0} + S_{1,1}, M_{1,0} + w, M_{0,1} + w \right]$$
$$= Max \left[0 + (-1), 0 + (-1), 0 + (-1) \right]$$
$$= Max \left[-1, -1, -1, \right]$$
$$= -1$$

The obtained score -1 is placed in position i, j (1,1) of the scoring matrix. Similarly using the above equation and method, fill all the remaining rows and columns. Place the back pointers to the cell from where the maximum score is obtained, which are predecessors of the current cell.

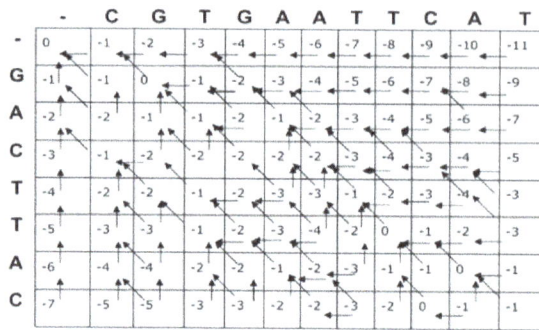

Matrix filling with back pointers

Trace back Step

The final step in the algorithm is the trace back for the best alignment. In the above mentioned example, one can see the bottom right hand corner score as -1. The important point to be noted here is that there may be two or more alignments possible between the two example sequences. The current cell with value -1 has immediate predecessor, where the maximum score obtained is diagonally located and its value is 0. If there are two or more values which points back, suggests that there can be two or more possible alignments.

By continuing the trace back step by the above defined method, one would reach to the 0th row, 0th column. Following the above described steps, alignment of two sample sequences can be found. The best alignment among the alignments can be identified by using the maximum alignment score (match =5, mismatch=-1, gap=-2) which may be user defined.

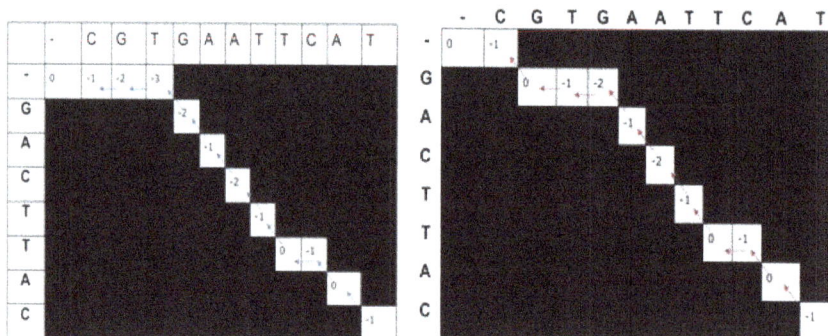

The possible Alignments with trace backing

Hirschberg's Algorithm

While the Needleman-Wunsch algorithm works well for sequence alignment, its space complexity (O(mn)) limits the size of sequences it can align. Hirschberg's algorithm uses a divide and conquer strategy to decrease the space requirement. Specifically, for two sequences (m and n) the first string is cut (m1 and m2) and the second string is cut in a corresponding place (into n1 and n2). The alignment is then solved recursively on m1 and n1, and m2 and n2. It is important to note that the two sub-strings (i.e. n1 and n2) need not have the same length.

Example:

With respect to sequence alignment, the objective is to take the sum at given intervals and use the value corresponding to the alignment score as the dividing point.

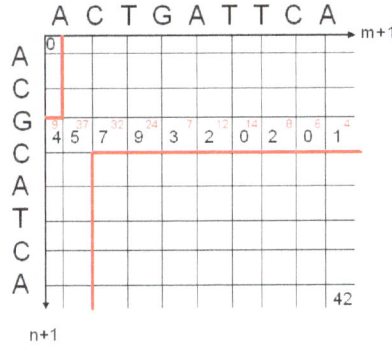

In this figure the sum up to the dividing line (37 + 5) adds up to the alignment score

Example:

What is the sequence of recursive calls on the following sequence alignment when using Hirschberg's algorithm:

    ACTG

    ACTT

The alignment is:

    s1: ACTG

        ||| s2: ACTT

Solution:

```
            s1[0..3] : s2[0..3]

              s1[0..1] : s2[0..1]

                s1[0..0] : s2[0..0]

                s1[1..1] : s2[1..1]

              s1[2..3] : s2[2..3]

                s1[2..2] : s2[2..2]

                s1[3..3] : s2[3..3]
```

Relating Local Alignments to Global Alignments

When aligning two very large sequences, it is often useful to determine the locations of high similarity regions. Now that we know how to calculate the *global* alignments, how can we find all local high-scoring hits, or *local* alignments above a given threshold for two large sequences? The answer is related to a programming "pearl", the 'Maximum Contiguous Subsequence Sum' (MSS).

Problem:

Given integers A_1, A_2, ..., A_N find (and identify the sequence corresponding to) the maximum value of:

$$\sum_{k=1}^{j} A_k$$

Solution:

Can be solved in time complexity of 'n'.

```
mss (A) {
    max = 0;
    sum = 0;
        for (i=1; i ≤ n; i+1)
            { sum = sum +
            A[i]; if (sum >
            max)
                    max =
            sum; if (sum <
            0)
                sum = 0;
    }
        return max;
    }
```

Analysis

When a subsequence occurs which has a negative sum, the subsequence which will be examined next can begin after the first subsequence (the one that produced the negative sum). Basically, the entire first subsequence is regarded as not having a starting point which will generate a positive sum. For example, consider this set of numbers:

4, 6, -2, 2, -14, 9

Some sums are positive $\left(4,\ 4+6,\ 4+6+(-2), 4+6+(-2)+2\right)$ but the sum of the first 5 terms $\left(4+6+(-2)+2-14\right)$ is negative. Therefore it follows logically that any sequence starting between the 4 and -14 and ending with the -14 will have a negative sum.

The maximum contiguous subsequence sum searches exactly for the highest scoring local area.

Smith–Waterman Algorithm

Over a decade after the initial publication of the Needleman-Wunsch algorithm, a modification was made to allow for local alignments. Today, the Smith-Waterman alignment algorithm is the one used by the Basic Local Alignment Search Tool (BLAST) which is the most cited resource in biomedical literature.

Working of Smith-Waterman Algorithm

Intialization of Matrix

The basic steps for the algorithm are similar to that of Needleman-Wunsch algorithm. The steps are:

1. Initialization of a matrix.

2. Matrix Filling with the appropriate scores.

3. Trace back the sequences for a suitable alignment.

To study the Local sequence alignment consider the given below sequences.

CGTGAATTCAT (sequence 1 or A)

GACTTAC (sequence 2 or B)

The two sequences are arranged in a matrix form with A+1columns and B+1rows. The values in the first row and first column are set to zero as shown in figure.

	-	C	G	T	G	A	A	T	T	C	A	T
-	0	0	0	0	0	0	0	0	0	0	0	0
G	0											
A	0											
C	0											
T	0											
T	0											
A	0											
C	0											

Initialization of Matrix

Variables used:

Variables used:

i, j describes row and columns.

M is the matrix value of the required cell (stated as $M_{i,j}$)

S is the score of the required cell $\left(S_{i,j}\right)$

W is the gap alignment

Matrix Filling

The second and crucial step of the algorithm is filling the entire matrix, so it is more important to know the neighbor values (diagonal, upper and left) of the current cell to fill each and every cell.

$$M_{i,j} = Maximum \left[M_{i-1,j-1} + S_{i,j}, M_{i,j-1} + W, M_{i-1,j} + W, 0 \right]$$

As per the assumptions stated earlier, fill the entire matrix using the assumed scoring schema and initial values. One can fill the 1st row and 1st column with the scoring matrix as follows.

The first residue (nucleotides or amino acids) in both sequences is 'C' and 'G', the matching score or the mismatching score is going to be added the neighboring value which is diagonally located i.e. 0. The upper and left values are added to the gap penalty score from the matrix. So the scoring schema equation can be shown as follows.

$$M_{1,1} = Maximum \left[M_{0,0} + S_{1,1}, M_{1,0} + w, M_{0,1} + W, 0 \right]$$
$$= Maximum \left[0\,(-3), 0 + (-4), 0 + (-4), 0 \right]$$
$$= Maximum \left[-3, -4, -4, 0 \right]$$
$$= 0$$

From the above calculations the maximum value obtained is 0. Finding the maximum value for $M_{i,j}$ position, one can notice that there is no chance to see any negative values in the matrix, since we are taking 0 as lowest value.

After filling the matrix, keep the pointer back to the cell from where the maximum score has been determined. In the similar fashion fill all the values of the matrix of the cell.

For the example the matrix can be filled is shown in figure.

	-	C	G	T	G	A	A	T	T	C	A	T
-	0	0	0	0	0	0	0	0	0	0	0	0
G	0	0	5	1	5	1	0	0	0	0	0	0
A	0	0	1	2	1	10	6	2	0	0	5	1
C	0	5	1	0	0	6	7	3	0	5	1	2
T	0	1	2	6	2	2	3	12	8	4	2	6
T	0	0	0	7	3	0	0	8	17	13	9	7
A	0	0	0	3	4	8	5	4	13	14	18	14
C	0	5	1	0	0	4	5	2	9	18	14	15

Matrix filling with back pointers

Each cell is back pointed by one or more pointers from where the maximum score has been obtained.

Trace Backing the Sequences for an Optimal Alignment

The final step for the appropriate alignment is trace backing, prior to that one needs to find out the maximum score obtained in the entire matrix for the local alignment of the sequences. It is possible that the maximum scores can be present in more than one cell, in that case there may be possibility of two or more alignments, and the best alignment by scoring it.

In this example we can see the maximum score in the matrix as 18, which is found in two positions that lead to multiple alignments, so the best alignment, has to be found.

So the trace back begins from the position which has the highest value, pointing back with the pointers, thus find out the possible predecessor, then move to next predecessor and continue until we reach the score 0.

Trace back of first possible alignment

It is possible to find two pointers pointing out from one cell, where both ways (alignments) can be considered, best one is found by scoring and finding maximum score among them.

Trace back of second possible alignment

Thus a local alignment is obtained and one can see the possible alignments as in figure.

```
G A A T T C A        G A A T T - C
| | | | |   |        | | | | |   |
G A C T T - A        G A C T T A C


+ + - + + - +        + + - + + - +
5 5 3 5 5 4 5        5 5 3 5 5 4 5
```

Scoring for best alignment

The two alignments can be given with a score, for matching as +5, mismatch as -3 and gap penalty as -4, sum up all the individual scores and the alignment which has maximum score after this can be taken as the best alignment.

By summing up the scores both of the alignments are giving the same as 18, so one can predict both alignments are the best.

Sequence Clustering

In bioinformatics, sequence clustering algorithms have been used to automatically group large protein datasets into different families, to search for protein sequences that are homologous to a given sequence, and to map or align a given DNA sequence to an entire genome, to cite only some of the most common applications. In all of these applications, sequence clustering becomes a valuable tool to gain insight into otherwise seemingly senseless sequences of data.

A similar kind of challenge arises in process mining, where the goal is to extract meaningful task sequences from an event log, usually resorting to special-purpose algorithms that can recover the original workflow that produced the log.

The idea of applying sequence clustering to process mining comes at a time when process mining is still heavily dependent on the assumption that the event log contains "sufficient" information, i.e., that each event in the log is clearly associated with a specific activity and case (process instance) . This comes as a major disadvantage since the classes of information systems that are able to generate such logs are restricted to process-aware systems, and it becomes impossible to apply and benefit from process mining in scenarios where the log data is not available in that form.

A sequence clustering approach can alleviate these requirements by grouping similar sequences and identifying typical ones without the need to provide any input information about the business logic. Of course, the results will bear a degree of uncertainty, whereas process mining approaches typically aim at finding exact models. Still, sequence clustering can provide valuable insight into the kind of sequences that are being executed.

Process Mining Approaches

In general, all process mining approaches take an event log as input and as a starting point for the discovery of underlying processes. The event log (also called process trace or audit trail) is list of records resulting from the execution of some process. For the log to be "minable", each record usually contains information about the activity that was executed, the process instance that it belongs to, and the time of execution. The requirements on the log, i.e. the kind of information it should contain, varies according to the process mining algorithm being used.

In fact, it is the choice of mining algorithms that often leads to different process mining approaches. Some of the algorithms used for process mining include:

- the α-algorithm: an algorithm that is able to re-create the Petri-net workflow from the ordering relations found in the even log. For the algorithm to work the log must contain the process instance identifier (case id) and it must be rather complete in the sense that all ordering relations should be present in the log.

- inference methods: a set of three different algorithms used to infer a finite state machine

(FSM) from an event log, where the log is regarded as a simple sequence of symbols. The three algorithms represent different levels of compromise between accuracy and robustness to noise. The MARKOV algorithm, inspired by Markov models, seems to be the most promising. The algorithm works by building up an event graph as the result of considering Markov chains with increasing order. In the last step, the graph is converted to a FSM, which represents the process that was found.

- directed acyclic graphs: an algorithm that is able to generate a dependency graph from a workflow system log. The log must contain a relatively high number of executions of the same process so that the dependency graph for that process can be completely built. Originally, the algorithm was proposed to support the adoption of workflow systems rather than actually pursuing process mining.

- inductive workflow acquisition: an approach in which the goal is to find a hidden markov model (HMM) that best represents the structure of the original process. The HMM can be found by either top-down or bottom-up refinement of an initial HMM structure; these are known as model splitting and model merging algorithms, respectively. The initial HMM structure is built directly from the log, which is regarded as a simple sequence of symbols. Reported results suggest that model splitting is faster and more accurate than model merging.

- *hierarchical clustering:* an algorithm that, given a large set of execution traces of a single process, separates them into clusters and finds the dependency graph separately for each cluster. The clusters of workflow traces are organized into a tree, hence the concept of model hierarchy. After the workflow models for the different clusters have been found, a bottom-up pass through the tree generalizes them into a single one.

- *genetic algorithm:* an algorithm in which several candidate solutions are evaluated by a fitness function that determines how consistent each solution is with the log. Every solution is represented by a causal matrix, i.e. a map of the input and output dependencies for each activity. Candidate solutions are generated by selection, crossover and mutation as in typical genetic algorithms. The search space is the set of all possible solutions with different combinations of the activities that appear in the event log. The log should contain a relatively high number of execution traces.

- *Instance graphs:* an approach that aims at portraying graphical representations of process execution, especially using Event-driven Process Chains (EPCs). For each execution trace found in the log, an instance graph is obtained for that process instance. In order to identify possible parallelism, each instance graph is constructed using the dependencies found in the entire log. Several instance graphs can then be aggregated in order to obtain the overall model for that log.

In general, as far as input data is concerned, all these algorithms require an event log that contains several, if not a very large number, of execution traces of the same process instance. (An exception is the RNET algorithm used in [8] which can receive a single trace as training input, but the results can vary widely depending on that given input sequence.) Because the log usually contains the traces of multiple instances, it is also required to have labelling field – usually called the *case id* – which specifies the process instance for every recorded event.

Another requirement on the content of the event log is that, for algorithms such as and, which rely on finding causal relations in the log, task A can be considered the cause of task B only if B follows A but A never follows B in the log. Exceptional behavior, errors or special conditions that would make A appear after B could ruin the results. These conditions are referred to as *noise*; algorithms that are able to withstand noise are said to be robust to noise. Most algorithms can become robust to noise by discarding causal relations with probability below a given threshold; this threshold is usually one of the algorithm parameters.

The problem with these requirements is that they may be difficult to apply in many potential scenarios for process mining. For example, in some applications the *case id* may be unavailable if the log is just an unclassified stream of recorded events. In other applications, it may be useful to clearly identify and distinguish normal behavior from exceptional one, without ruling out small variations simply as noise. These issues suggest that other kind of algorithms could provide valuable insight into the original behavior that produced the log. If there is no *case id* available, and there is an unpredictable amount of ad-hoc behavior, then an algorithm that allows us to sort out and understand that behavior could be the first step before actually mining those processes. Sequence clustering algorithms are a good candidate for this job.

Sequence Clustering

Sequence clustering is a collection of methods that aim at partitioning a number of sequences into meaningful clusters or groups of similar sequences. The development of such methods has been an active field of research especially in connection with challenges in bioinformatics. Here we will present the basic principles by referring to a simple sequence clustering algorithm based on first-order Markov chains.

In this algorithm, each cluster is associated with a first-order Markov chain, where the current state depends only on the previous state. The probability that an observed sequence belongs to a given cluster is in effect the probability that the observed sequence was produced by the Markov chain associated with that cluster. For a sequence $x = \{x_0, x_1, x_2, ..., x_{L-1}\}$ of length L this can be expressed simply as:

$$p(x|c_k) = p(x_0, c_k) \cdot \prod_{i=1}^{i-L-1} p(x_i | x_{i-1}, c_k)$$

where $p(x_0, c_k)$ is the probability of x_0 occurring as the first state in the Markov chain associated with cluster c_k and $p(x_i | x_{i-1}, c_k)$ is the transition probability of state x_{i-1} to state xi in that same Markov chain. Given the way to compute $p(x|c_k)$, the sequence clustering algorithm can be implemented as an extension to the well-known Expectation-Maximization (EM) algorithm. The steps are:

1. Initialize the model parameters $p(x, c)$ and $p(x_i | x_{i-1}, c_k)$ randomly, i.e. for each cluster the state transition probabilities of the associated Markov chain are initialized at random.

2. Using the current model parameters, assign each sequence to each cluster with a probability given by equation $p(x|c_k) = p(x_0, ck) \cdot \prod_{i=1}^{i-L-1} p(x_i | x_{i-1}, c_k)$.

3. Use the results of step 2 to re-estimate the model parameters, i.e. recalculate the state transition probabilities of each Markov chain based on the sequences that belong to that cluster.

4. Repeat steps 2 and 3 until the mixture model converges.

The algorithm must be provided with two input tables: a case table and a nested table. The case table contains one record for each sequence; it conveys the number of sequences in the input data set together with some descriptive information about each sequence. The nested table contains the steps for all sequences, where each step is numbered and labelled. The number is the order of occurrence within the sequence, and the label is a descriptive attribute that denotes the state in a Markov chain. The case and nested tables share a one-to-many relationship: each sequence in the case table is associated with several steps in the nested table. The connecting attribute, which is application-specific, serves as key in the case table and as sequence scope delimiter in the nested table.

For the sake of clarity, let us consider a simple example. Suppose the members of a given family have different ways of zapping through TV channels according to their own interests. Let us assume that each member always finds the TV switched off, and after turning it on, goes through a set of channels before turning it off again. Every time it is turned on, the TV generates a new session identifier (session id) and records both session-related information as well as the sequence of channel changes. Figure below shows the case and nested tables for this scenario. The session identifier is both the key to the case table and the sequence scope delimiter for the nested table. The case table contains descriptive, non-sequence attributes about each session, whereas the nested table contains the steps for each sequence, both numbered and labelled.

sessionid	timeofday		sessionid	channel	numchange
1234	2007-04-09 19:41:00		1234	Music	1
1235	2007-04-09 21:15:00		1234	Music	2
1236	2007-04-10 18:56:00		1234	Movies	3
1237	2007-04-10 20:38:00		1234	Movies	4
1238	2007-04-10 21:25:00		1235	News	1
1239	2007-04-11 19:06:00		1235	Sports	2
1240	2007-04-11 20:23:00		1235	News	3

Case (a) and nested (b) tables for the simple TV usage scenario

It can be seen from this simple example that the input data to be provided to the sequence clustering algorithm already has a form of *case id*, which is the session identifier. Pre-processing techniques will have to be used to assign this *case id* if it is not available in the first place.

What is interesting to note here is the kind of results that the sequence clustering algorithm is able to produce. Figure shows four of the clusters that the algorithm was able to identify from a given set of 24 sequences for the simple TV usage scenario. Each cluster has a different Markov chain that is able to generate the sequences assigned to that cluster. This effectively captures the dominant behavior of similar sequences.

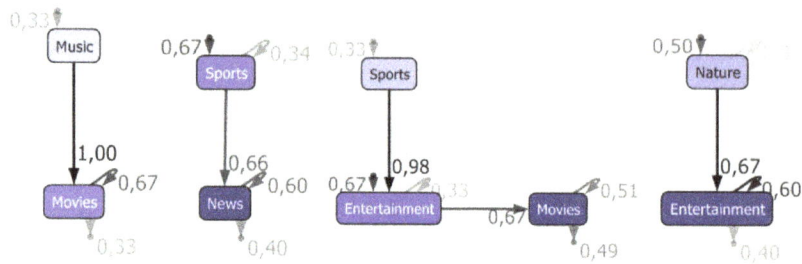

The Markov chains in four of the clusters obtained for the simple TV usage scenario.

The number of clusters to be found can be set manually or automatically by letting the algorithm perform a heuristic to determine the number of clusters for the given data. This is usually very useful to use as an initial guess before trying to run the algorithm with different parameters.

To produce the results shown in figure the algorithm performed a number of iterations, where each iteration comprises two steps: the *expectation step* and the *maximization step*. In the expectation step the algorithm assigns each sequence x to the cluster ck that gives the highest membership probability $p(x|c_k)$. Once this step is complete, the algorithm has a provisional estimate of which sequences belong to which cluster. In the maximization step the algorithm re-computes the transition probabilities $p(x_i|x_{i-1},c_k)$ for each cluster c_k based on the sequences that belong to that cluster. After the maximization step, the next expectation step will produce different results from the previous iteration, since $p(x|c_k)$ will now be computed with the updated values of $p(x_i|x_{i-1},c_k)$. The algorithm converges when there is no change in the values of these model parameters.

Pairwise Algorithm

Pairwise sequence alignment is a challenging task because of the exponential growth of genomic information, necessitating large-scale comparison of two input strings. The size of GenBank/the European Molecular Biology Laboratory (EMBL)/the DNA DataBank of Japan (DDBJ) nucleotide database is now doubling in every 15 months. To search databases to find out sequences similar to a given query sequence, the search programs compute an alignment score for every sequence in the database. This score represents the degree of similarity between the query and database sequence. A dynamic programming algorithm for computing the optimal local alignment score was first described by Smith and Waterman, and later improved in for linear gap penalty functions. Though dynamic programming is the best alignment procedure so far procedure so far, but it is not suitable for large strings in terms of both time and space. For two strings of lengths m and n, the time and space complexities of the Smith and Waterman (SW) algorithm are $O(mn)$ The time and space complexity had been improved to O (rn) in, where r is the amount of allowed error, by considering only the useful part of the distance matrix. However, for large error rates, r is $O(m)$, so the complexity is still $O(mn)$. Later on, the space complexity of SW was improved to $O(n)$. Dynamic programming has been accelerated through GLASS by first finding exactly matching long substrings, but the time and space complexity are still high. LAGAN is another implementation of dynamic programming, but is not applicable on a genome scale without prior information ("anchors") that directs comparison to orthologous regions.

Pairwise Alignment

Of the many proposed methods for analyzing biological networks, global network alignment is one of the most ambitious. We are given two graphs $G_1 = (V_1, E_1)$ and $G_2 = (V_2, E_2)$, whose vertices represent proteins, and the presence of the edge (u, v) in $E_1 (E_2)$ indicates that the two proteins represented by u and v interact in $G_1 (G_2)$. Most aligners assume, without loss of generality, that $|V_1| < |V_2|$.

The problem of pairwise alignment is to find a one-to-one function $f : V_1 \rightarrow V_2$ that maps each node in V_1 to the node in V_2 that it best matches [as a shorthand to make several equations more readable, we will treat f as a function on edges as well. In this case, $f((u,v))$ is simply more readable shorthand for $(f(u), f(v))$. It must be noted that some algorithms produce a *partial* function, abstaining from mapping nodes that cannot be matched well.

Most aligners decompose the process of producing a matching into two steps. First, for each pair of nodes in $V_1 \times V_2$, we compute their similarity by examining the local topology of the graph around those two nodes and by their sequence similarity, as measured by BLAST bit scores or E-values. Second, taking the similarities between these nodes as weighted edges in a bipartite graph with two sets of nodes V_1, V_2, we solve the maximum-weight bipartite matching problem to generate a mapping from V_1 to V_2. The pairwise alignment software available differs primarily in how they handle these two steps, with most innovation being focused on the first step of estimating the topological similarity of two nodes. This process is just a general schematic that applies to many aligners. Some, such as NATALIE 2.0, do not follow the two-step process and instead optimize a relaxed version of the problem directly.

We must stress that all existing aligners only *approximately* solve this problem, and they generally introduce approximations in two ways. First, they introduce a relaxed problem definition. Second, they use a heuristic algorithm to approximately solve the relaxed problem. For instance, GRAAL frames the problem as matching nodes to one another in such a way as to maximize their graphlet signature similarity (a measure of how many small subgraphs of various shapes are in both nodes' respective neighborhoods), in the hopes that maximizing this metric between all aligned pairs of nodes will produce a biologically informative alignment. Then, a greedy matching heuristic is used that aligns the most similar pair of nodes first and then works outward, aligning their neighborhoods. When evaluating this algorithm, it is not clear how much of its performance is attributable to the metric of graphlet signature similarity, and how much is due to using a matching algorithm that prefers to map neighbors to neighbors. All we can do is evaluate how well the heuristic solution to the relaxed problem solves our original problem. If a given aligner performs poorly, it could be that the relaxed problem is a good choice, but that choice is hamstrung by a poorly designed heuristic to solve it. Without being able to swap out the parts of these aligners, however, all we can do is evaluate their final results.

Genetic Algorithm

Genetic Algorithm (GA) is a search-based optimization technique based on the principles of Genetics and Natural Selection. It is frequently used to find optimal or near-optimal solutions to difficult problems which otherwise would take a lifetime to solve. It is frequently used to solve optimization problems, in research, and in machine learning.

Optimization

Optimization is the process of making something better. In any process, we have a set of inputs and a set of outputs as shown in the following figure.

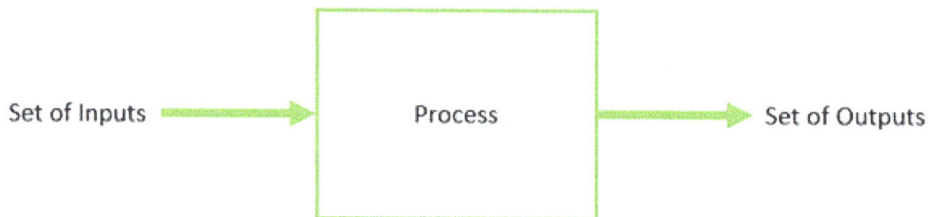

Optimization refers to finding the values of inputs in such a way that we get the "best" output values. The definition of "best" varies from problem to problem, but in mathematical terms, it refers to maximizing or minimizing one or more objective functions, by varying the input parameters.

The set of all possible solutions or values which the inputs can take make up the search space. In this search space, lies a point or a set of points which gives the optimal solution. The aim of optimization is to find that point or set of points in the search space.

Nature has always been a great source of inspiration to all mankind. Genetic Algorithms (GAs) are search based algorithms based on the concepts of natural selection and genetics. GAs are a subset of a much larger branch of computation known as Evolutionary Computation.

GAs were developed by John Holland and his students and colleagues at the University of Michigan, most notably David E. Goldberg and has since been tried on various optimization problems with a high degree of success.

In GAs, we have a pool or a population of possible solutions to the given problem. These solutions then undergo recombination and mutation (like in natural genetics), producing new children, and the process is repeated over various generations. Each individual (or candidate solution) is assigned a fitness value (based on its objective function value) and the fitter individuals are given a higher chance to mate and yield more "fitter" individuals. This is in line with the Darwinian Theory of "Survival of the Fittest".

In this way we keep "evolving" better individuals or solutions over generations, till we reach a stopping criterion.

Genetic Algorithms are sufficiently randomized in nature, but they perform much better than random local search (in which we just try various random solutions, keeping track of the best so far), as they exploit historical information as well.

Advantages of GAs

GAs has various advantages which have made them immensely popular. These include:

- Does not require any derivative information (which may not be available for many re-al-world problems).

- Is faster and more efficient as compared to the traditional methods.

- Has very good parallel capabilities.

- Optimizes both continuous and discrete functions and also multi-objective problems.

- Provides a list of "good" solutions and not just a single solution.

- Always gets an answer to the problem, which gets better over the time.

- Useful when the search space is very large and there are a large number of parameters involved.

Limitations of GAs

Like any technique, GAs also suffer from a few limitations. These include:

- GAs are not suited for all problems, especially problems which are simple and for which derivative information is available.

- Fitness value is calculated repeatedly which might be computationally expensive for some problems.

- Being stochastic, there are no guarantees on the optimality or the quality of the solution.

- If not implemented properly, the GA may not converge to the optimal solution.

Generic Structure of Genetic Algorithm

- Population: It is a subset of all the possible (encoded) solutions to the given problem. The population for a GA is analogous to the population for human beings except that instead of human beings, we have Candidate Solutions representing human beings.

- Chromosomes: A chromosome is one such solution to the given problem.

- Gene: A gene is one element position of a chromosome.

- Allele: It is the value a gene takes for a particular chromosome.

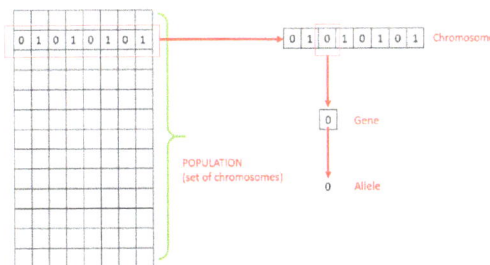

- Genotype: Genotype is the population in the computation space. In the computation space, the solutions are represented in a way which can be easily understood and manipulated using a computing system.

- Phenotype: Phenotype is the population in the actual real world solution space in which solutions are represented in a way they are represented in real world situations.

- Decoding and Encoding: For simple problems, the phenotype and genotype spaces are the same. However, in most of the cases, the phenotype and genotype spaces are different. Decoding is a process of transforming a solution from the genotype to the phenotype space, while encoding is a process of transforming from the phenotype to genotype space. Decoding should be fast as it is carried out repeatedly in a GA during the fitness value calculation.

For example, consider the 0/1 Knapsack Problem. The Phenotype space consists of solutions which just contain the item numbers of the items to be picked.

However, in the genotype space it can be represented as a binary string of length n (where n is the number of items). A 0 at position x represents that x^{th} item is picked while a 1 represents the reverse. This is a case where genotype and phenotype spaces are different.

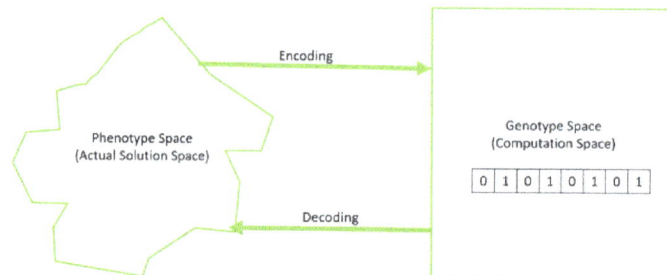

- Fitness Function: A fitness function simply defined is a function which takes the solution as input and produces the suitability of the solution as the output. In some cases, the fitness function and the objective function may be the same, while in others it might be different based on the problem.

- Genetic Operators: These alter the genetic composition of the offspring. These include crossover, mutation, selection, etc.

The basic structure of a GA is as follows:

We start with an initial population (which may be generated at random or seeded by other heuristics), select parents from this population for mating. Apply crossover and mutation operators on the parents to generate new off-springs. And finally these off-springs replace the existing individuals in the population and the process repeats. In this way genetic algorithms actually try to mimic the human evolution to some extent.

Selection

In principle, a population of individuals selected from the search space, often in a random manner, serves as candidate solutions to optimize the problem. The individuals in this population are evaluated through ("fitness") adaptation function. A selection mechanism is then used to select individuals to be used as parents to those of the next generation. These individuals will then be crossed and mutated to form the new offspring. The next generation is finally formed by an alternative mechanism between parents and their offspring. This process is repeated until a certain satisfaction condition.

Six different selection methods are considered here, namely: the roulette wheel selection (RWS), the stochastic universal sampling (SUS), the linear rank selection (LRS), the exponential rank selection (ERS), the tournament selection (TOS), and the truncation selection (TRS).

Roulette Wheel Selection (RWS)

The conspicuous characteristic of this selection method is the fact that it gives to each individual i of the current population a probability $p(i)$ of being selected, proportional to its fitness $f(i)$.

$$p(i) = \frac{f(i)}{\sum_{j=1}^{n} f(j)}$$

Where n denotes the population size in terms of the number of individuals. The RWS can be implemented according to the following pseudo-code

RWS_ pseudo-code

{

- Calculate the sum $S = \sum_{i=1}^{n} f(j)$

- For each individual do { li n $\leq i \leq n$ do {

  - Generate a random number $\alpha \in [0, S]$
  - $i\ Sum = 0;\quad j = 0;$

  - Do {
  - $iSum \leftarrow iSum + f(j);$
  - $j \leftarrow +1;$
  } while ($iSum < \alpha$ and $j < n$)
    - Select the individual j ;}

}

Note that a well-known drawback of this technique is the risk of premature convergence of the GA to a local optimum, due to the possible presence of a dominant individual that always wins the competition and is selected as a parent.

Stochastic Universal Sampling (SUS)

The SUS is a variant of RWS aimed at reducing the risk of premature convergence. It can be implemented according to the following pseudo-code:

```
SUS_ pseudo-code
```

{

- Calculate the mean $\bar{f}=1/n\sum_{i=1}^{n}f(j)$;
- Generate a random number $\alpha\in[0,1]$;
- $Sum=f(1)$; $delta=\alpha\times\bar{f}$; $j=0$;
- Do {

        • If (delta < Sum) {

                - select the jth individual;

                -   delta = delta + Sum ;

            }

            else {

                -    $j=j+1$;

                -    $Sum=Sum+f(j)$;

            }

        } while $(j<n)$

}

Linear Rank Selection (LRS)

LRS is also a variant of RWS that tries to overcome the drawback of premature convergence of the GA to a local optimum. It is based on the rank of individuals rather than on their fitness. The rank n is accorded to the best individual whilst the worst individual gets the rank 1. Thus, based on its rank, each individual *i* has the probability of being selected given by the expression

$$p(i)=\frac{rank(i)}{n\times(n-1)}$$

Once all individuals of the current population are ranked, the LRS procedure can be implemented according to the following pseudo-code:

```
LRS_ pseudo-code
```

    {

Calculate the sum $v=\dfrac{1}{n-2.001}$;

- For each individual $1 \leq i \leq n$ do {

  - Generate a random num $\alpha \in [0, v]$;

    - For each $1 \leq j \leq n$ do {

      - If ($p(j) \leq \alpha$) {

        - Select the jth individual;

      - Break;

      }

    }

  }

}

Exponential Rank Selection (ERS)

The ERS is based on the same principle as LRS, but it differs from LRS by the probability of selecting each individual. For ERS, this probability is given by the expression:

$$p(i)1.0*\exp\left(\frac{-rang\,(i)}{c}\right)$$

With

$$c = \frac{(n*2*(n-1))}{(6*(n-1)+n)}$$

Once the n probabilities are computed, the rest of the method can be described by the following pseudo-code:

ERS_ pseudo-code

{

- For each individual $1 \leq j \leq n$ do {

  - Generate a random number $\alpha \in \left[\frac{1}{9}c, \frac{2}{c}\right]$;

  - For

  - each $1 \leq j \leq n$ do {

      - If ($p(j) \leq \alpha$) {

```
                         - Select the jth individual;

                              - Break;

                         } // end if

                    } // end for j

              }// end for i

}
```

Tournament Selection (TOS)

Tournament selection is a variant of rank-based selection methods. Its principle consists in randomly selecting a set of k individuals. These individuals are then ranked according to their relative fitness and the fittest individual is selected for reproduction. The whole process is repeated times for the entire population. Hence, the probability of each individual to be selected is given by the expression:

$$p(i) \begin{cases} \dfrac{C_{n-1}^{k-1}}{C_n^k} \text{ if } i \in [1, n-k-1] \\ 0 \text{ if } i \in [n-k, n] \end{cases}$$

Technically speaking, the implementation of TOS can be performed according to the pseudo-code:

```
    TOS_ pseudo-code

    {

    •  Create a table  t  where the  n   individuals are placed in a randomly
       chosen order

    •  For  i = 1 to n do {

    •  for  j = 1 to n do {

        •  i1 = t( j) ;

        •  For m = 1 to n do {

            •  i2 = t( j + m) ;

            •  If ( f (i1 ) > f (i2 ) ) the select i1

            }// end for m

        •  j = j + k ;

        } // end for j


        }// end for i

}
```

Truncation Selection (ERS)

The truncation selection is a very simple technique that orders the candidate solutions of each population according to their fitness. Then, only a certain portion p of the fittest individuals are selected and reproduced 1/ p times. It is less used in practice than other techniques, except for very large population. The pseudo-code of the technique is as follows:

```
TRS_ pseudo-code

{

•   Order the n  individuals of fitness;

•   Set the portion p  of individuals to select (e.g.10% ≤ p ≤ 50%);

•   sp = int(n × p) // selection pressure;


•   Select the first sp  individuals;


}
```

Crossover

The crossover operator is a genetic operator that combines (mates) two chromosomes (parents) to produce a new chromosome (offspring). The idea behind crossover is that the new chromosome may be better than both of the parents if it takes the best characteristics from each of the parents. Crossover occurs during evolution according to a user definable crossover probability.

Single Point Crossover

When performing crossover, both parental chromosomes are split at a randomly determined crossover point. Subsequently, a new child genotype is created by appending the first part of the first parent with the second part of the second parent. A single crossover point on both parents' organism strings is selected. All data beyond that point in either organism string is swapped between the two parent organisms. Figure shows the single point crossover (SPC) process.

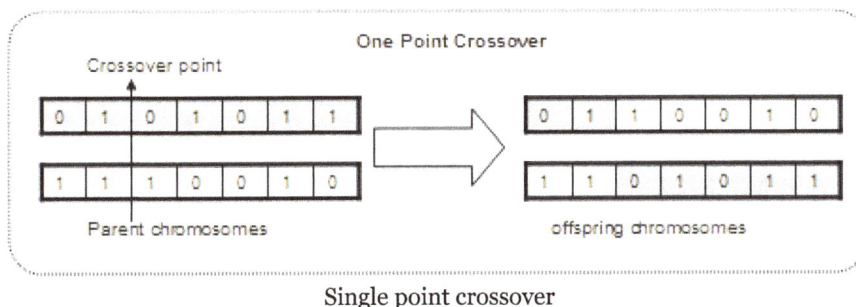

Single point crossover

Two Point Crossover

Apart from SPC, many different crossover algorithms have been devised, often involving more than one cut point. It should be noted that adding further crossover points reduces the performance of the GA. The problem with adding additional crossover points is that building blocks are more likely to be disrupted. However, an advantage of having more crossover points is that the problem space may be searched more thoroughly. In two-point crossover (TPC), two crossover points are chosen and the contents between these points are exchanged between two mated parents.

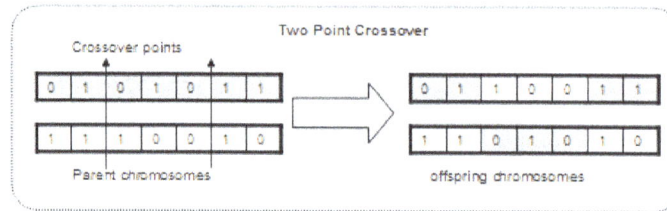

Two point crossover

In figure above, the arrows indicate the crossover points. Thus, the contents between these points are exchanged between the parents to produce new children for mating in the next generation.

Intermediate Crossover

Intermediate creates offsprings by a weighted average of the parents. Intermediate crossover (IC) is controlled by a single parameter Ratio:

$$offspring = parent1 + rand * Ratio * (parent2 - parent1)$$

If Ratio is in the range then the offsprings produced are within the hypercube defined by the parents locations at opposite vertices. Ratio can be a scalar or a vector of length number of variables. If Ratio is a scalar, then all of the offsprings will lie on the line between the parents. If Ratio is a vector then children can be any point within the hypercube.

Heuristic Crossover

In heuristic crossover (HC), heuristic returns an offspring that lies on the line containing the two parents, a small distance away from the parent with the better fitness value in the direction away from the parent with the worse fitness value. The default value of Ratio is 1.2. If parent1 and parent2 are the parents, and parent1 has the better fitness value, the function returns the child,

$$offspring = parent2 + Ratio * (parent1 - parent2)$$

Arithmetic Crossover

In arithmetic crossover (AC), arithmetic creates children that are the weighted arithmetic mean of two parents. Children are feasible with respect to linear constraints and bounds. Alpha is random value between. If parent1 and parent2 are the parents, and parent1 has the better fitness value, the function returns the child,

$$offspring = alpha * parent1 + (1 - alpha) * parent2$$

Mutation

In simple terms, mutation may be defined as a small random tweak in the chromosome, to get a new solution. It is used to maintain and introduce diversity in the genetic population and is usually applied with a low probability – p_m. If the probability is very high, the GA gets reduced to a random search.

Mutation is the part of the GA which is related to the "exploration" of the search space. It has been observed that mutation is essential to the convergence of the GA while crossover is not.

Mutation Operators

We describe some of the most commonly used mutation operators. This is not an exhaustive list and the GA designer might find a combination of these approaches or a problem-specific mutation operator more useful.

Bit Flip Mutation

In this bit flip mutation, we select one or more random bits and flip them. This is used for binary encoded GAs.

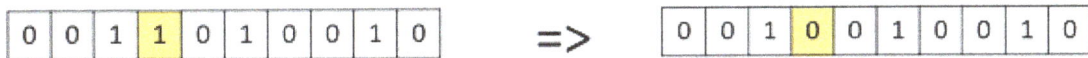

0	0	1	1	0	1	0	0	1	0	=>	0	0	1	0	0	1	0	0	1	0

Random Resetting

Random Resetting is an extension of the bit flip for the integer representation. In this, a random value from the set of permissible values is assigned to a randomly chosen gene.

Swap Mutation

In swap mutation, we select two positions on the chromosome at random, and interchange the values. This is common in permutation based encodings.

1	2	3	4	5	6	7	8	9	0	=>	1	6	3	4	5	2	7	8	9	0

Scramble Mutation

Scramble mutation is also popular with permutation representations. In this, from the entire chromosome, a subset of genes is chosen and their values are scrambled or shuffled randomly.

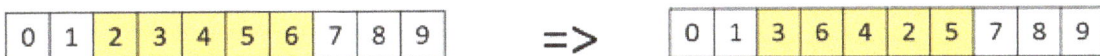

0	1	2	3	4	5	6	7	8	9	=>	0	1	3	6	4	2	5	7	8	9

Inversion Mutation

In inversion mutation, we select a subset of genes like in scramble mutation, but instead of shuffling the subset, we merely invert the entire string in the subset.

Schema

The theoretical background of genetic algorithms is seen most clearly using the binary string representation and the classic algorithm. First the wildcard $\star$ shall match any single component of a vector. A schema is built of the alphabet of genes (the fixed components) augmented by the wildcard symbol, e.g., $(0,1,\star)$ matches $(0,1,0)$ and $(0,1,1)$ if the alphabet is $\{0,1\}$. Hence a schema represents all strings of the search space which match the schema at all components except $\star$.

The following analysis is performed for the classic genetic algorithm, whose inner loop is the following.

1. $t := t+1.$

2. Select the new population $P(t)$ from the old one $P(t-1)$ by single string selection based on a positive fitness function.

3. Recombine $P(t)$ by one point crossover and mutation.

4. Evaluate $P(t)$.

5. Repeat.

Let $\delta(S)$ denote the defining length of a schema , i.e., the distance between the first and the last fixed component. It measures how compact the information contained in a schema is. For example $\delta\big((\star,\star,0,1,\star,1,0)\big)=4$. Let $o(S)$ denote the order of the schema S, i.e., the number of fixed components. For example $o\big((\star,\star,0,1,\star,1,0)\big)=4$. Let $\mathrm{eval}(S,t)$ be the fitness of the schema S at time t, i.e., the average fitness of all strings in the population that are matched by S. If S matches the n individuals $\{v_1,...,v_n\}$, then $\mathrm{eval}(S,t):=(1/k)\sum_{k=1}^{n}\mathrm{eval}(v_k)$, where $\mathrm{eval}(v_k)$ is the fitness of a single individual v_k. Finally $F(t)$ is the total fitness of the whole population at time t, i.e., $F(t):=\sum_k \mathrm{eval}(v_k)$. $N(t)$ is the size of the population at time t.

In the single string selection step of the classic genetic algorithm considered, an individual v_k has the probability $\mathrm{eval}(v_k)/F(t)$ of being selected. After the selection step it is expected that

$$\xi(S,t+1)=\xi(S,t)\cdot N(t)\cdot \frac{\mathrm{eval}(S,t)}{F(t)}$$

strings match the schema S, since S matches $\xi(S,t)$ individuals, the number of single string selections is $N(t)$, and the probability that an average string matched by S is selected in a single string selection is $\mathrm{eval}(S,t)/F(t)$.

Introducing the average fitness of the population $\bar{F}(t):=F(t)/N(t)$, this can be written as

$$\xi(S,t+1)=\xi(S,t)\cdot \frac{\mathrm{eval}(S,t)}{\bar{F}(t)}.$$

The interpretation of this formula is that the number of individuals matching a schema S grows in each time step like the ratio of the fitness of S and the total fitness. Thus an above average schema, i.e., a schema with $\text{eval}(S,t) > \overline{F}(t)$, will proliferate successfully in the next generation, but a below average schema will not. In the long term the formula implies that above average schemata will proliferate exponentially.

The next step after selection is recombination, i.e., crossover and mutation, where new individuals are introduced into the population. The effects of crossover are considered first. A crossover site is selected uniformly among $m-1$ possible sites, where m is the length of the vector representing an individual. Hence $\delta(S)/(m-1)$ is the probability that the schema S is destroyed if a certain vector (or chromosome) undergoes crossover, and thus

$$p_{cs}(S) \geq 1 - p_c \frac{\delta(S)}{m-1}$$

holds for the probability $p_{cs}(S)$ of its survival, where p_c is the probability of crossover. $p_{cs}(S)$ is greater equal, and not equal, to the right hand side, since it may happen that a schema is conserved after splitting it in middle if both ancestors happen to contain parts of the schema at the right places.

Thus taking into account selection and crossover, the number of individuals matching a schema S can be estimated by

$$\xi(S,t+1) \geq \xi(S,t) \cdot \frac{\text{eval}(S,t)}{\overline{F}(t)} \cdot \left(1 - p_c \frac{\delta(S)}{m-1}\right).$$

The final operator to be considered is mutation. Let p_m be the probability that a single component is changed. Since mutations are independent from one another, the probability that a schema S survives a mutation is

$$p_{ms}(S) = (1 - p_m)^{o(S)}.$$

Furthermore $p_{ms} \approx 1 - o(S)p_m$, since $p_m \ll 1$.

Hence the final estimation of $(S,t \ 1)$ is

$$\xi(S,t+1) \gtrsim \xi(S,t) \cdot \frac{\text{eval}(S,t)}{\overline{F}(t)} \cdot \left(1 - p_c \frac{(S)}{m \ 1} - o(S)p_m\right),$$

which takes selection, crossover, and mutation in the classic genetic algorithm into account.

The Schema theorem can thus be stated as follows.

The inequality

$$\xi(S,t+1) \gtrsim \xi(S,t) \cdot \frac{\text{eval}(S,t)}{\overline{F}(t)} \cdot \left(1 - p_c \frac{\delta(S)}{m-1} - o(S)p_m\right),$$

holds with the above notation. This means that short, low order, above average schemata receive exponentially increasing trials in subsequent generations of the classic genetic algorithm and below average schemata receive exponentially decreasing trials.

Chromosome

A very common problem in adaptive control is learning to balance a pole that is hinged on a cart that can move in one dimension along a track of fixed length, as show in Figure . The control must use `bang-bang' control, that is, a force of fixed magnitude can be applied to push the cart to the left or right.

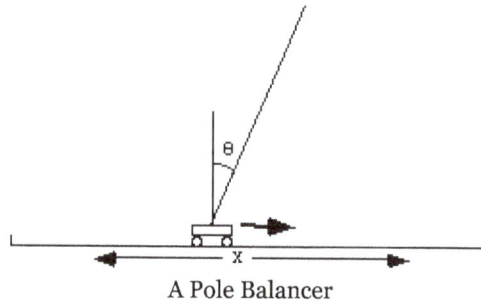

A Pole Balancer

Before we can begin to learn how to control this system, it is necessary to represent it somehow. We will use the BOXES method that was devised by Michie and Chambers (1968). The measurements taken of the physical system are the angle of the pole, θ and its angular velocity and the position of the cart, x, and its velocity. Rather than treat the four variables as continuous values, Michie and Chambers chose to discretize each dimension of the state space. One possible discretization is shown in figure.

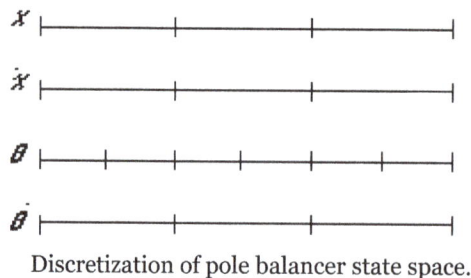

Discretization of pole balancer state space.

This discretization results in 3 x 3 x 6 x 3 = 162 `boxes' that partition the state space. Each box has associated with it an action setting which tells the controller that when the system is in that part of the state space, the controller should apply that action, which is a push to the left or a push to the right. Since there is a simple binary choice and there are 162 boxes, there are 2^{162} possible control strategies for the pole balancer.

The simplest kind of learning in this case, is to exhaustively search for the right combination. However, this is clearly impractical given the size of the search space. Instead, we can invoke a genetic search strategy that will reduce the amount of search considerably.

In genetic learning, we assume that there is a population of individuals, each one of which, represents a candidate problem solver for a given task. Like evolution, genetic algorithms test each individual from the population and only the fittest survive to reproduce for the next generation. The algorithm creates new generations until at least one individual is found that can solve the problem adequately.

Each problem solver is a *chromosome*. A position, or set of positions in a chromosome is called a *gene*. The possible values (from a fixed set of symbols) of a gene are known as *alleles*. In most genetic algorithm implementations the set of symbols is $\{0, 1\}$ and chromosome lengths are fixed. Most implementations also use fixed population sizes.

The most critical problem in applying a genetic algorithm is in finding a suitable encoding of the examples in the problem domain to a chromosome. A good choice of representation will make the search easy by limiting the search space, a poor choice will result in a large search space. For our pole balancing example, we will use a very simple encoding. A chromosome is a string of 162 boxes. Each box, or gene, can take values: 0 (meaning push left) or 1 (meaning push right). Choosing the size of the population can be tricky since a small population size provides an insufficient sample size over the space of solutions for a problem and large population requires a lot of evaluation and will be slow. In this example, 50 is a suitable population size.

Each iteration in a genetic algorithm is called a *generation*. Each chromosome in a population is used to solve a problem. Its performance is evaluated and the chromosome is given some rating of fitness. The population is also given an overall fitness rating based on the performance of its members. The fitness value indicates how close a chromosome or population is to the required solution. For pole balancing, the fitness value of a chromosome may be the number of time steps that the chromosome is able to keep the pole balanced for.

New sets of chromosomes are produced from one generation to the next. Reproduction takes place when selected chromosomes from one generation are recombined with others to form chromosomes for the next generation. The new ones are called *offspring*. Selection of chromosomes for reproduction is based on their fitness values. The average fitness of population may also be calculated at end of each generation. For pole balancing, individuals whose fitness is below average are replaced by reproduction of above average chromosomes. The strategy must be modified if two few or too many chromosomes survive. For example, at least 10% and at most 60% must survive.

Operators that recombine the selected chromosomes are called genetic operators. Two common operators are *crossover* and *mutation*. Crossover exchanges portions of a pair of chromosomes at a randomly chosen point called the crossover point. Some Implementations have more than one crossover point. For example, if there are two chromosomes, X and Y:

$$X = 1001\ 01011 \qquad Y = 1110\ 10010$$

and the crossover point is 4, the resulting offspring are:

$$O1 = 100110010 \qquad O2 = 1110\ 01011$$

Offspring produced by crossover cannot contain information that is not already in the population, so an additional operator, mutation, is required. Mutation generates an offspring by randomly changing the values of genes at one or more gene positions of a selected chromosome. For example, if the following chromosome,

$$Z = 100101011$$

is mutated at positions 2, 4 and 9, then the resulting offspring is:

$$O = 110001010$$

The number of offspring produced for each new generation depends on how members are introduced so as to maintain a fixed population size. In a *pure* replacement strategy, the whole population is replaced by a new one. In an *elitist* strategy, a proportion of the population survives to the next generation.

In pole balancing, all offspring are created by crossover (except when more the 60% will survive for more than three generations when the rate is reduced to only 0.75 being produced by crossover). Mutation is a background operator which helps to sustain exploration. Each offspring produced by crossover has a probability of 0.01 of being mutated before it enters the population. If more than 60% will survive, the mutation rate is increased to 0.25.

The number of offspring an individual can produce by crossover is proportional to its fitness:

$$\frac{\text{fitness value}}{\text{population fitness}} \times \text{No. children}$$

where the number of children is the total number of individuals to be replaced. Mates are chosen at random among the survivors.

Fitness Function

Fitness Function (also known as the Evaluation Function) evaluates how close a given solution is to the optimum solution of the desired problem. It determines how fit a solution is.

In genetic algorithms, each solution is generally represented as a string of binary numbers, known as a chromosome. We have to test these solutions and come up with the best set of solutions to solve a given problem. Each solution, therefore, needs to be awarded a score, to indicate how close it came to meeting the overall specification of the desired solution. This score is generated by applying the fitness function to the test, or results obtained from the tested solution.

Generic Requirements of a Fitness Function

The following requirements should be satisfied by any fitness function.

1. The fitness function should be clearly defined. The reader should be able to clearly understand how the fitness score is calculated.

2. The fitness function should be implemented efficiently. If the fitness function becomes the bottleneck of the algorithm, then the overall efficiency of the genetic algorithm will be reduced.

3. The fitness function should quantitatively measure how fit a given solution is in solving the problem.

4. The fitness function should generate intuitive results. The best/worst candidates should have best/worst score values.

Fitness Function for a Given Problem

Each problem has its own fitness function. The fitness function that should be used depends on the given problem. Coming up with a fitness function for the given problem is the hardest part when it comes to formulating a problem using genetic algorithms.

There is no hard and fast rule that a particular function should be used in a particular problem. However, certain functions have been adopted by data scientists regarding certain types of problems.

Typically, for classification tasks where supervised learning is used, error measures such as *Euclidean distance* and *Manhattan distance* have been widely used as the fitness function.

For optimization problems, basic functions such as sum of a set of calculated parameters related to the problem domain can be used as the fitness function.

Let's go through a few example problems and their related fitness functions.

Example — Generating Sequences

Given a set of 5 genes, which can hold one of the binary values 0 and 1, we have to come up with the sequence having all 1s. So we have to maximize the number of 1s as much as possible. This can be considered as an optimization problem. Hence, the fitness function is considered as the *number of 1s present in the genome*. If there are five 1s, then it is having maximum fitness and solves our problem. If there are no 1s, then if has the minimum fitness. The code given below shows how the fitness function is being implemented to calculate the fitness score.

```
class
Individual
{
            int fitness = 0;
            int geneLength = 5;
            int[] genes = new int[5];

            public Individual() {
              Random rn = new Random();
             //Set genes randomly for each individual
              for (int i = 0; i < geneLength; i++) {
                genes[i] = rn.nextInt() % 2;
              }
              fitness = 0;
            }
            //Calculate fitness
            public void calculateFitness() {
              fitness = 0;
              for (int i = 0; i < geneLength; i++) {
                if (genes[i] == 1) {
                  ++fitness;
                }
              }
            }
          }
```

Genetic Operator

Genetic operators are used in genetic algorithms to generate diversity (*mutation*-like operators) and to combine existing solutions into others (*crossover*-like operators). The main difference among them is that the former operate on one chromosome, that is, they are *unary*, while the latter are binary operators. In GAGS, genetic operators need not be coerced into the mutation/crossover paradigm, they are just functions that act on chromosomes.

In GAGS, operators constitute a class, that is, they are not methods of the chromosome class as is usual. That means that they can be created, destroyed and changed in runtime, and besides, that they can be subclassed to create new operators. This class is designed to act as a *functor*, i.e., function-syntax objects, which have operator ( ) overloaded; this is only syntactic sugar, but allows to approach the looks of the C++ implementation to the actual algorithm.

They act in two different ways: non-directed, which means that they act on a random part of the chromosome, or directed, that is, they act on a preselected part of the chromosome. Besides, many operators consider the chromosome is divided in chunks, usually called *genes*; the size of those chunks must be pased as a parameter to their constructors.

Some operators are variable-length operators: they alter the length of the chromosome; chromosome length is always computed in every method that needs it, which means that GAGS is prepared for variable-length chromosomes. Length is always change by a discrete amount of genes. Note also that binary operators acting on variable length chromosomes can also change the length of the resulting chromosome.

An example that includes binary and unary operators is as follows

```
#include <genop.hpp>        // Chromosomes already included


main () {

  const unsigned NUMGENES = 4;

  const bitLength_t SIZEGENES = 3;

  seed_random( time( (time_t) 0 ) );// From randcl

  chrom aChromosome( NUMGENES*SIZEGENES ); // Create chromosome

  cout << "aChromosome\t\t\t" << aChromosome << endl;

  genOp creeper( SIZEGENES, genOp::CREEP ); // Apply unary genOp

  creeper( &aChromosome );

  cout << "aChromosome creeped\t\t" << aChromosome << endl;

  genOp mutator( (mutRate_t) 0.1 ); // Another unary genOp

  mutator( &aChromosome );

  cout << "aChromosome mutated 0.1\t" << aChromosome << endl;
```

```
genOp SGA2pt( (unsigned char) 2 ); // Binary genOp, 2-point crossover

bitString aBS ="111111111111"; //

chrom anotherChrom( aBS ); // Define other Chromosome from string

cout << "Crossovering with anotherChrom\n\t\t\t"

    << aChromosome << "\n\t\t\t" << anotherChrom << endl ;

SGA2pt( &aChromosome, &anotherChrom );

cout << "Result\t\t" << aChromosome << endl;

}
```

This program defines two unary operators: creeper, which changes a gene by plus or minus one and mutate, which does the usual thing; and a binary operator, SGA2pt, a simple two point cross-over, and applies them to the chromosomes defined. A gene has been defined as a 3-bit segment, and each chromosome has got 4 genes. Output would look like this:

```
aChromosome             100011011110

aChromosome creeped     100010011110

aChromosome mutated 0.1 100010011010

Crossovering with anotherChrom

             100010011010

             111111111111

Result       100010011110
```

Al The genetic operator type is computed in two different ways: if its constructor has got unique parameters, the genetic operator type is deduced; if not, typed enums are used.

genOp::DUP	duplicates a gene with mutation
genOp::KILL	eliminates a gene
genOp::RANDINC	adds a random gene
genOp::TRANSP	transposes two genes, i.e., permutates its contents
genOp::CLONE	copies the chromosome without doint anything
genOp::MUT	usual bit-flip mutation
genOp::UXOVER	uniform crossover: interchanges bits between the two chromosomes
genOp::XOVER	usual n-point crossover: interchanges the part of the chromosome between two or more randomly generated points
genOp::GXOVER	same as before, but respects gene boundaries, which means that only whole genes are interchanged
genOp::ZAP	changes the value of a gene or part of a chromosome to another value
genOp::CREEP	changes the value of a gene by plus or minus one.

Genetic operators and modes.

Using the modes shown in table, genetic operators use three different constructors:

```
genOp( mutRate_t _rate,

    genOpMode _mode = MUT);   // Mutation and uniform crossover

genOp( bitLength_t _lenBits,

    genOpMode _mode = TRANSP); // Most operators acting on genes

genOp( unsigned char _numPts ); // number of xOver points > 1
```

Genetic operators can also be applied in a directed way, by using the applyAt method

```
const unsigned toKill = 1;

chrom thisChrom( "111000111000");

genOp killer( SIZEGENES, genOp::KILL);

killer.applyAt( toKill*SIZEGENES, &thisChrom );

cout << "Result\t\t\t" << thisChrom << endl;
```

BLAST

As gene and protein sequence databases grew at the end of the twentieth century, scientists turned to computers to help analyze this abundant and ever-growing amount of data. Today, one of the most common tools used to examine DNA and protein sequences is the Basic Local Alignment Search Tool, also known as BLAST. BLAST is a computer algorithm that is available for use online at the National Center for Biotechnology Information (NCBI) website, as well as many other sites. BLAST can rapidly align and compare a query DNA sequence with a database of sequences, which makes it a critical tool in ongoing genomic research. In recent years, the parallel development of large-scale sequencing projects and bioinformatic tools like BLAST has enabled scientists to study the genetic blueprint of life across many species, and it has also helped connect biology and computer science in the maturing field of bioinformatics.

Alignment Theory

Although the computer science principles behind BLAST have been around for some time, prior to BLAST, they had not been applied to biology. Before BLAST, alignment programs used dynamic programming algorithms, such as the Needleman-Wunsch and Smith-Waterman algorithms, that required long processing times and the use of supercomputers or parallel computer processors.

Figure depicts a Needleman-Wunsch alignment of the words "PELICAN" and "COELACANTH." The search space of the alignment is shown using a Cartesian grid and is proportional to the length of the sequences being compared plus one extra row and column.

Initialization of the alignment matrix.

A Needleman-Wunsch alignment of the words "PELICAN" and "COELACANTH."

Next, the alignment matrix is initialized with a zero in the upper left corner. For each letter of the word being aligned, a point is deducted so that each letter has a progressively more negative score. Why does the algorithm subtract a point? In an alignment, the diagonal is read from the upper left to the lower right, and when the analysis moves vertically or horizontally, it indicates a gap in the sequence. Thus, each time the program moves straight up or down, a gap penalty is applied that takes away points from the alignment score. Finally, a little arrow, or pointer, is added to indicate which direction to follow the alignment. In the third stage, the algorithm starts to actually build and score the alignment in a step called fill, or induction.

Filling the axes of the alignment matrix.

When filling the axes of the alignment matrix, start in the upper left corner and set it to 0. Next, assign a score for each letter in the row or column. Note that there is a penalty for gaps, and that the arrow should point toward the origin of the alignment.

In this example, the analysis begins by aligning the C to the P and calculating a score. In figure, one point is added if two letters match, and one point is subtracted if they do not. This calculation is carried out three times, once for each square to the left (dark blue), above (green), and to the upper left (brown). Using a value from either the upper square or the left square, the final score is -2 (-1 + -1). Using the 0 score in the upper left diagonal square, the final score is -1 (0 + -1). Because -1 is the highest score, this score is jotted down in the alignment matrix, and because the upper left square was the one leading to the best score, an arrow is inserted in the box pointing toward this square.

Induction or filling of the alignment matrix, part I.

One point is added if two letters match, and one point is subtracted if they do not. Using a value from either the upper (green) or left (dark blue) square, the final score is -2; however, using the value from the upper left (brown) square, the final score is -1. Because this is the highest score, it is recorded in the alignment matrix along with an arrow pointing to the upper left square.

This same process continues, calculating two scores for every square in the matrix. At the end, there is a completely scored matrix with a series of arrows used to find the optimal alignment.

Induction or filling in of the alignment matrix, part II.

The same process is carried out for the next square in the alignment. Here, using the value in upper left (brown) square yields a sum of -2, using the value in the upper (green) square yields a sum of -3, and using the value in the left (dark blue) square yields a sum of -2. Because -2 is the highest score and was initially calculated using the upper left square, -2 is recorded in the matrix along with an arrow pointing toward the brown square.

Induction or filling in of the alignment matrix, part III.

The rest of the matrix is completed using the same method.

The final steps of generating an alignment are called traceback, and they involve finding the optimal, highest-scoring alignment. The traceback starts in the lower right of the matrix and follows the pointers to the adjacent boxes. By definition, this will be the best scoring path through the alignment.

Traceback of the optimal alignment.

The final steps of generating an alignment are called traceback, which involved finding the optimal, highest-scoring alignment.

Traceback of the optimal complete alignment.

During traceback, following the best scoring path reveals the best alignment.

Although this sort of dynamic programming did a complete job of comparing every single residue of one sequence to every single residue of a second sequence and kept track of how well the sequences aligned at every step, these algorithms required a considerable amount of computer memory and processing time. Computing speed was an especially important concern, because these exhaustive programs had to search databases that continued to grow at exponential rates. Moreover, most regions of the search space did not score very well and therefore probably could have been skipped during the calculation process. Finally, these programs required powerful computing hardware that was expensive, rare, and ultimately impractical for most scientists and labs.

Researcher Stephen Altschul and colleagues wanted to bypass these challenges and develop a way for databases to be searched quickly on routinely used computers. In order to increase the speed of alignment, the BLAST algorithm was designed to approximate the results of an alignment algorithm created by Smith and Waterman, but to do so without comparing each residue against every other residue. BLAST is therefore heuristic in nature, meaning it has "smart shortcuts" that allow it to run more quickly. However, in this trade-off for increased speed, the accuracy of the algorithm is slightly decreased.

BLAST Heuristic

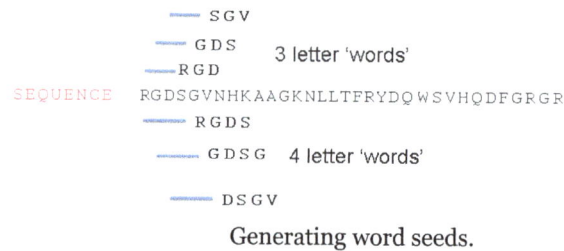

Generating word seeds.

Instead of comparing every residue against each other, BLAST uses short "word" (w) segments to create alignment "seeds." BLAST is designed to create a word list from the query sequence with words of a specific length, as defined by the user.

BLAST increases the speed of alignment by decreasing the search space or number of comparisons it makes. Specifically, instead of comparing every residue against each other, BLAST uses short "word" (w) segments to create alignment "seeds." BLAST is designed to create a word list from the query sequence with words of a specific length, as defined by the user. Requiring only three residues to match in order to seed an alignment means that fewer sequence regions need to be compared. Larger word sizes usually mean that there are even fewer regions to evaluate. Once an alignment is seeded, BLAST extends the alignment according to a threshold (T) that is set by the user. When performing a BLAST query, the computer extends words with a neighborhood score greater than T. A cutoff score (S) is used to select alignments over the cutoff, which means the sequences share significant homologies. If a hit is detected, then the algorithm checks whether w is contained within a longer aligned segment pair that has a cutoff score greater than or equal to S. When an alignment score starts to decrease past a lower threshold score (X), the alignment is terminated. These and many other variables can be adjusted to either increase the speed of the algorithm or emphasize its sensitivity.

Testing the BLAST Algorithm

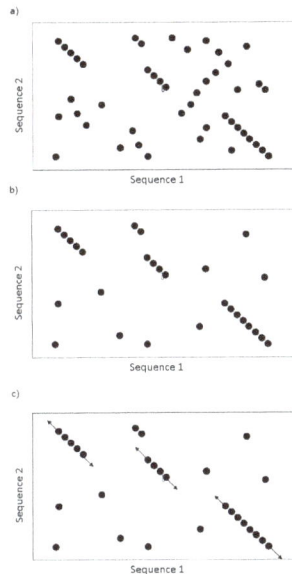

We tested the BLAST algorithm on a database of randomly generated sequences, and we examined the output resulting from different w and T parameters. If T is set to be a lower threshold, then the algorithm detects more word pairs and requires a longer processing time . Thus, choosing the value for T was a major decision because the researchers wanted to reach a compromise between the algorithm's sensitivity and its processing time.

Next, we tested BLAST on a database of real sequences, and they found it was successful in quickly identifying alignments with high scores. In searching the globin gene family, for example, we found that BLAST identified 88 of the 89 globin alignments that scored above 80. Other gene families, including the immunoglobulins, protein kinases, and cytochrome c genes, were then examined to measure the number of alignments detected when using different T and S values. BLAST was also able to detect similar regions within pairs of long sequences. These tests therefore showed that BLAST was fast, sensitive, and accurate as a tool for analyzing sequence alignments.

Bringing Mathematical Rigor to Alignment

One of the most notable innovations of BLAST is that the program calculates the statistical significance for each sequence alignment result. This is known as the expect value (E-value) or probability value (P-value), and it is calculated for each alignment. The E-value describes how many hits you can expect to see by chance when searching a database of a certain size, whereas the P-value describes the probability that the alignment you are observing is due to chance. In general, the lower the E- or P-value is, the more likely it is that an alignment is significant. Below the common 10^{-5} score, P and E are roughly equivalent.

The addition of statistical rigor to sequence alignment has been controversial. Some researchers rely too much on significance values to include or exclude sequences despite poorly chosen parameters, whereas others over-interpret "insignificant" results because the results "look" right. While all scientific results are subject to interpretation, BLAST scores and statistics bring much-needed objectivity to sequence comparisons, and the debate about them has helped improve methods for determining significance.

BLAST Family

Since 1990, many variants of BLAST have been developed, each with specialized features. Early on, the original BLAST was split into two adaptations: NCBI BLAST and Washington University BLAST (WU BLAST). Both BLASTs have program variations. For instance, BLASTN can be used to compare a nucleotide sequence with a nucleotide database; BLASTP can be used to compare a protein sequence with a database of protein sequences; and BLASTX can take a nucleotide sequence, translate it, and query it versus a protein database in one step. TBLASTN compares a protein query sequence to all six possible reading frames of a database and is often used to identify proteins in new, undescribed genomes. Finally, TBLASTX compares all six reading frames of a query sequence to all six reading frames of a database-an intensive algorithmic feat that can bring even modern computers to a grinding halt if not used properly.

In addition, NCBI has some of its own specialized variants of BLAST. For example, MEGABLAST is a program that can rapidly complete searches for sequences with only minor variations and can more efficiently manage queries with longer sequences. PSI- and PHI- are other powerful BLAST tools that allow more complex and evolutionary divergent proteins to be aligned.

Thus, since its creation, BLAST has become an essential tool for biologists. Its speed and sensitivity allow scientists to compare nucleotide and protein sequences to both single sequences and large databases. Most importantly, BLAST has helped democratize bioinformatics analysis and make it accessible to any researcher over the Internet; indeed, it is rare to read a modern molecular biology paper that does not refer to a BLAST alignment. In short, BLAST and its descendant applications have permitted scientists to predict the functions of genes and proteins in whole genomes, answering questions *in silico* that could never be answered at a lab bench or in the field.

BLAT

Analyzing vertebrate genomes requires rapid mRNA/DNA and cross-species protein alignments. A new tool, BLAT, is more accurate and 500 times faster than popular existing tools for mRNA/DNA alignments and 50 times faster for protein alignments at sensitivity settings typically used when comparing vertebrate sequences. BLAT's speed stems from an index of all nonoverlapping K-mers in the genome. This index fits inside the RAM of inexpensive computers, and need only be computed once for each genome assembly. BLAT has several major stages. It uses the index to find regions in the genome likely to be homologous to the query sequence. It performs an alignment between homologous regions. It stitches together these aligned regions (often exons) into larger alignments (typically genes). Finally, BLAT revisits small internal exons possibly missed at the first stage and adjusts large gap boundaries that have canonical splice sites where feasible.

BLAT, which is short for "BLAST-like alignment tool." BLAT is similar in many ways to BLAST. The program rapidly scans for relatively short matches (hits), and extends these into high-scoring pairs (HSPs). However, BLAT differs from BLAST in some significant ways. Where BLAST builds an index of the query sequence and then scans linearly through the database, BLAT builds an index of the database and then scans linearly through the query sequence. Where BLAST triggers an extension when one or two hits occur in proximity to each other, BLAT can trigger extensions on any number of perfect or near-perfect hits. Where BLAST returns each area of homology between two sequences as separate alignments, BLAT stitches them together into a larger alignment. BLAT has special code to handle introns in RNA/DNA alignments. Therefore, whereas BLAST delivers a list of exons sorted by exon size, with alignments extending slightly beyond the edge of each exon, BLAT effectively "unsplices" mRNA onto the genome—giving a single alignment that uses each base of the mRNA only once, and which correctly positions splice sites.

BLAT is used to produce the human EST and mRNA alignments. The human EST alignments compared 1.75×10^9 bases in 3.73×10^6 ESTs against 2.88×10^9 bases of human DNA and took 220 CPU hours on a Linux farm of 800 MhZ Pentium IIIs. BLAT was used in translated mode to align a 2.5× coverage unassembled whole-genome shotgun of the mouse versus the masked human genome. This involved 7.51×10^9 bases in 1.33×10^7 reads and took 16,300 CPU hours. The client/server version of BLAT is used to power untranslated and translated interactive searches on http://genome.ucsc.edu. Researchers all over the world use BLAT to perform thousands of interactive sequence searches per day. The nucleotide server has sustained over 500,000 search requests per day from program-driven queries. We do ask those researchers who are doing more than a few thousand program-driven queries to obtain a copy of BLAT to use on their own servers. The nucleotide server is not as efficient as the stand-alone program, since to save memory it does not keep the genome in memory, only the index. The index uses approximately 1 gigabyte on unmasked DNA in untranslated mode, and approximately 2.5 gigabytes on masked DNA in translated mode. The translated mode server by default is less sensitive than the default stand-alone settings. It requires three perfect amino acid 4-mers to trigger an alignment. The untranslated server usually responds to a 1000-base cDNA query in less than a second. The translated server usually responds to a 400-amino acid protein query in <5 sec.

Methods

Algorithm

All fast alignment programs break the alignment problem into two parts. Initially in a "search stage," the program detects regions of the two sequences which are likely to be homologous. The program then in an "alignment stage" examines these regions in more detail and produces alignments for the regions which are indeed homologous according to some criteria. The goal of the search stage is to detect the vast majority of homologous regions while reducing the amount of sequence that is passed to the alignment stage.

Searching with Single Perfect Matches

A simple and reasonably effective search stage is to look for subsequences of a certain size, k, which are shared by the query sequence and the database. In many practical implementations of this search, every K-mer in the query is compared against all nonoverlapping K-mers in the database. Let's examine the number of homologous regions that are missed, and the number of nonhomologous regions that are passed to the alignment stage using these criteria. First, we'll need some definitions:

K: The K-mer size. Typically this is 8–16 for nucleotide comparisons and 3–7 for amino acid comparisons.

M: The match ratio between homologous areas. This would be typically about 98% for cDNA/genomic alignments within the same species, about 89% for protein alignments between human and mouse.

H: The size of a homologous area. For a human exon this is typically 50–200 bases.

G: The size of the database—3 billion bases for the human genome.

Q: The size of the query sequence.

A: The alphabet size; 20 for amino acids, 4 for nucleotides.

Assuming that each letter is independent of the previous letter, the probability that a specific K-mer in a homologous region of the database matches perfectly the corresponding K-mer in the query is simply:

$$p_1 = M^k$$

It is convenient to introduce a term that counts the number of nonoverlapping K-mers in the homologous region:

$$T = \mathrm{floor}\left(H/K\right)$$

The probability that at least one nonoverlapping K-mer in the homologous region matches perfectly with the corresponding K-mer in the query is:

$$P = 1 - \left(1 - p_1\right)^T = 1 - \left(1 - M^K\right)^T$$

The number of nonoverlapping K-mers that are expected to match by chance, assuming that all letters are equally likely to occur is:

$$F = \left(Q - K + 1\right)^* \left(G/K\right)^* \left(1/A\right)^K$$

Tables below show P and F values for various levels of sequence identity and K-mer sizes. For EST alignments we might want the search phase to find at least 99% of sequences that have 5% or less sequencing noise. Looking at Table below, to achieve this level of sensitivity using this simple search method, we would need to choose a K of 14 or less. A K of 14 results in 399 regions passed on to the alignment phase by chance alone. Any smaller K would pass significantly more. Mouse and human sequences average 89% identity at the amino acid level. Looking at Table below to compare a translated mouse read and find at least 99% of the sequences at this level of identity we would need a K of 5 or less, which would result in 62,625 sequences passed on to the alignment stage. Depending on the cost of the alignment stage, these simple search criteria may or may not be suitable. Comparing mouse and human coding sequences at the nucleotide level, where there is on average 86% base identity, requires us to reduce our K to 7 to find at least 99% of the sequences. This results in 13,078,962 regions passed to the alignment stage, which would probably not be practical.

Sensitivity and Specificity of Single Perfect Nucleotide K-mer Matches as a Search Criterion

	7	8	9	10	11	12	13	14
A. 81%	0.974	0.915	0.833	0.726	0.607	0.486	0.373	0.314
83%	0.988	0.953	0.897	0.815	0.711	0.595	0.478	0.415
85%	0.996	0.978	0.945	0.888	0.808	0.707	0.594	0.532
87%	0.999	0.992	0.975	0.942	0.888	0.811	0.714	0.659
89%	1.000	0.998	0.991	0.976	0.946	0.897	0.824	0.782
91%	1.000	1.000	0.998	0.993	0.981	0.956	0.912	0.886
93%	1.000	1.000	1.000	0.999	0.995	0.987	0.968	0.957
95%	1.000	1.000	1.000	1.000	0.999	0.998	0.994	0.991
97%	1.000	1.000	1.000	1.000	1.000	1.000	1.000	0.999
B. K	7	8	9	10	11	12	13	14
F	1.3e+07	2.9e+06	635783	143051	32512	7451	1719	399

(A) Columns are for K sizes of 7–14. Rows represent various percentage identities between the homologous sequences. The table entries show the fraction of homologies detected as calculated from equation 3 assuming a homologous region of 100 bases. The larger the value of K, the fewer homologies are detected.

(B) K represents the size of the perfect match. F shows how many perfect matches of this size expected to occur by chance according to equation 4 in a genome of 3 billion bases using a query of 500 bases.

Sensitivity and Specificity of Single Perfect Amino Acid K-mer Matches as a Search Criterion

K	3	4	5	6	7
A. 71%	0.992	0.904	0.697	0.496	0.317
73%	0.996	0.931	0.752	0.560	0.374
75%	0.998	0.952	0.803	0.625	0.436
77%	0.999	0.969	0.850	0.689	0.503
79%	0.999	0.981	0.890	0.752	0.574
81%	1.000	0.989	0.924	0.810	0.646
83%	1.000	0.994	0.950	0.862	0.718
85%	1.000	0.997	0.970	0.906	0.787
87%	1.000	0.999	0.984	0.942	0.850
89%	1.000	1.000	0.993	0.968	0.903
91%	1.000	1.000	0.997	0.985	0.945
93%	1.000	1.000	0.999	0.995	0.975
B. K	3	4	5	6	7
F	4.2e+07	1.6e+06	62625	2609	112

(A) Columns are for K sizes of 3–7. Rows represent various percentage identities between the homologous sequences. The table entries show the fraction of homologies detected as calculated from equation 3 assuming a homologous region of 33 amino acids. (B) K represents the size of the perfect match. F shows how many perfect matches of this size are expected to occur by chance according to equation 4 in a translated genome of 3 billion bases using a query of 167 amino acids (corresponding to 500 bases).

Searching with Single almost Perfect Matches

What if instead of requiring perfect matches with a K-mer to trigger an alignment, we allow almost perfect matches, that is, hits where one letter may mismatch? The probability that a nonoverlapping K-mer in a homologous region of the database matches almost perfectly the corresponding K-mer in the query is:

$$p_1 = K^* M^{k-1} \, {}^* (1 - M) + M^K$$

As with a single perfect hit, the probability that any nonoverlapping K-mer in the homologous region matches almost perfectly with the corresponding K-mer in the query is:

$$P = 1 - (1 - p_1)^T$$

Whereas the number of K-mers which match almost perfectly by chance are:

$$F = (Q - K + 1)^* \, (G/K)^* \left(K^* (1/A)^{K-1*} (1 - (1/A)) + (1/A)^K \right)$$

Tables below show P and F for various levels of sequence identity and K-mer sizes. For the purposes of EST alignments, a K of 22 or less would pass through over 99% of the truly homologous regions while on average passing less than one chance match through to the aligner. With a reasonably fast alignment stage, it would be feasible to look for mouse/human homologies at the nucleotide level using this technique. A K size of 12 detects over 99% of the mouse homologies, and requires checking 275,671 alignments. At the amino acid level, a K size of 8 has the desired sensitivity and requires checking only 374 alignments.

Sensitivity and Specificity of Single Near-Perfect (One Mismatch Allowed) Nucleotide K-mer Matches as a Search Criterion

	12	13	14	15	16	17	18	19	20	21	22
A. 81%	0.945	0.880	0.831	0.721	0.657	0.526	0.465	0.408	0.356	0.255	0.218
83%	0.975	0.936	0.904	0.820	0.770	0.649	0.591	0.535	0.480	0.361	0.318
85%	0.991	0.971	0.954	0.900	0.865	0.767	0.719	0.669	0.619	0.490	0.445
87%	0.997	0.990	0.983	0.954	0.935	0.867	0.833	0.796	0.757	0.634	0.591
89%	1.000	0.997	0.995	0.984	0.976	0.939	0.920	0.897	0.872	0.775	0.741
91%	1.000	1.000	0.999	0.996	0.994	0.979	0.971	0.962	0.950	0.890	0.869
93%	1.000	1.000	1.000	0.999	0.999	0.996	0.994	0.991	0.988	0.963	0.954
95%	1.000	1.000	1.000	1.000	1.000	1.000	0.999	0.999	0.999	0.994	0.992
97%	1.000	1.000	1.000	1.000	1.000	1.000	1.000	1.000	1.000	1.000	1.000
B. K	12	13	14	15	16	17	18	19	20	21	22
F	275671	68775	17163	4284	1070	267	67	17	4.2	1.0	0.3

(A) Columns are for K sizes of 12–22. Rows represent various percentage identities between the homologous sequences. The table entries show the fraction of homologies detected as calculated by equation 6 assuming a homologous region of 100 bases. (B) K represents the size of the near-perfect match. F shows how many perfect matches of this size expected to occur by chance according to equation 7 in a genome of 3 billion bases using a query of 500 bases.

Sensitivity and Specificity of Single Near-Perfect (One Mismatch Allowed) Amino Acid K-mer Matches as a Search Criterion

	4	5	6	7	8	9
A. 71%	1.000	0.992	0.946	0.823	0.725	0.515
73%	1.000	0.995	0.965	0.867	0.785	0.586
75%	1.000	0.998	0.978	0.905	0.840	0.657
77%	1.000	0.999	0.987	0.935	0.886	0.727
79%	1.000	0.999	0.993	0.959	0.924	0.791
81%	1.000	1.000	0.997	0.976	0.952	0.849
83%	1.000	1.000	0.999	0.987	0.973	0.897
85%	1.000	1.000	0.999	0.994	0.986	0.936
87%	1.000	1.000	1.000	0.997	0.994	0.964
89%	1.000	1.000	1.000	0.999	0.998	0.982
91%	1.000	1.000	1.000	1.000	0.999	0.993
93%	1.000	1.000	1.000	1.000	1.000	0.998
B. K	4	5	6	7	8	9
F	1.2E+08	6.0E+06	300078	14985	749	37

(A) Columns are for K sizes of 4–9. Rows represent various percentage identities between the homologous sequences. The table entries show the fraction of homologies detected. (B) K represents the size of the near-perfect match. F shows how many perfect matches of this size expected to occur by chance in a translated genome of 3 billion bases using a query of 167 amino acids.

Searching with Multiple Perfect Matches

Another alternative search method is to require multiple perfect matches that are constrained to be near each other. Consider a situation where the K size is 10 and there are two hits—one starting at position 10 in the query and 1010 in the database, and another starting at position 30 in the query and 1030 in the database. These two hits could easily be part of a region of homology extending from positions 10–39 in the query and 1010–1039 in the database. If we subtract the query coordinate from the database coordinate, we get a "diagonal" coordinate. Consider the search criteria that there must be N perfect matches, each no further than W letters from each other in the target coordinate, and have the same diagonal coordinate. For N = 1, the probability that a nonoverlapping K-mer in a homologous region of the database matches perfectly the corresponding K-mer in the query is simply as before:

$$p_1 = M^K$$

The probability that there are exactly n matches within the homologous region is

$$p_n = p_1^n * (1 - p_1)^{T-n} * T! / (n! * (T - n)!)$$

And the probability that there are N or more matches is the sum:

$$P = P_N + P_{N+1} + \ldots + P_T$$

The number of sets of N perfect matches that occur by chance is a little complex to calculate. For $N = 1$ it is easy:

$$F_1 = (Q - K + 1)^* (G/K)^* (1/A)^K$$

The probability of a second match occuring within W letters after the first is

$$S = 1 - \left(1 - (1/A)^K\right)^{W/K}$$

because the second match can occur with any of the W/K nonoverlapping K-mers in the database within W letters after the first match. We can extend this reasoning to consider the chance that the Nth match is within W letters after the $(N-1)^{th}$ match, which gives the more general relationship

$$F_N = S * F_{N-1}$$

which can be solved as

$$F_N = F_1 * S^{N-1}$$

where F_N represents the number of chance matches of N K-mers each separated by no more than W from the previous match.

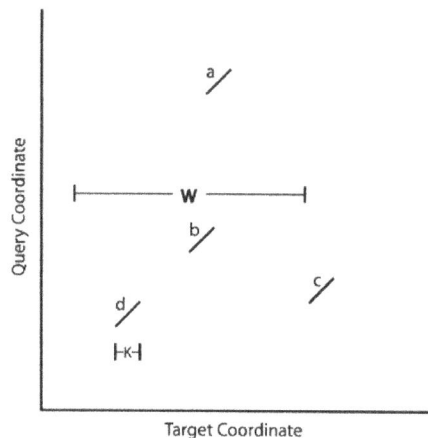

A pair of hits and two other hits. The hits a, b, c, and d are all K letters long. Hits d and b have the same diagonal coordinate and are within W letters of each other. Therefore they would match the "two perfect K-mer" search criteria.

Tables below show the sensitivity and specificity for N values of 2 and 3 and various values of other parameters which approximate cDNA or mouse/human alignments.

Sensitivity and Specificity of Multiple (2 and 3) Perfect Nucleotide K-mer Matches as a Search Criterion

	2,8	2,9	2,10	2,11	2,12	3,8	3,9	3,10	3,11	3,12
A. 81%	0.681	0.508	0.348	0.220	0.129	0.389	0.221	0.112	0.051	0.021
83%	0.790	0.638	0.475	0.326	0.208	0.529	0.339	0.193	0.099	0.045
85%	0.879	0.762	0.615	0.460	0.318	0.676	0.487	0.313	0.180	0.093
87%	0.942	0.866	0.752	0.611	0.461	0.809	0.649	0.470	0.305	0.177
89%	0.978	0.940	0.868	0.761	0.625	0.910	0.801	0.648	0.476	0.314
91%	0.994	0.980	0.947	0.884	0.787	0.969	0.914	0.815	0.673	0.505
93%	0.999	0.996	0.986	0.962	0.912	0.993	0.976	0.933	0.851	0.722
95%	1.000	1.000	0.998	0.993	0.979	0.999	0.997	0.987	0.961	0.902
97%	1.000	1.000	1.000	1.000	0.999	1.000	1.000	0.999	0.997	0.987
B. N,K	2,8	2,9	2,10	2,11	2,12	3,8	3,9	3,10	3,11	3,12
F	524	27	1.4	0.1	0.0	0.1	0.0	0.0	0.0	0.0

(A) Columns are for N sizes of 2 and 3 and K sizes of 8–12. Rows represent various percentage identities between the homologous sequences. The table entries show the fraction of homologies detected as calculated by equation 10. (B) N and K represent the number and size of the near-perfect matches, respectively. F shows how many perfect clustered matches expected to occur by chance according to equation 14 in a translated genome of 3 billion bases using a query of 167 amino acids.

Sensitivity and Specificity of Multiple (2 and 3) Perfect Amino Acid K-mer Matches as a Search Criterion

	2,3	2,4	2,5	2,6	2,7	3,3	3,4	3,5	3,6	3,7
A. 71%	0.945	0.643	0.297	0.126	0.044	0.945	0.643	0.297	0.126	0.044
73%	0.965	0.712	0.363	0.167	0.063	0.965	0.712	0.363	0.167	0.063
75%	0.978	0.776	0.436	0.218	0.089	0.978	0.776	0.436	0.218	0.089
77%	0.987	0.833	0.514	0.280	0.123	0.987	0.833	0.514	0.280	0.123
79%	0.993	0.882	0.596	0.353	0.169	0.993	0.882	0.596	0.353	0.169
81%	0.997	0.922	0.678	0.435	0.226	0.997	0.922	0.678	0.435	0.226
83%	0.999	0.952	0.757	0.526	0.298	0.999	0.952	0.757	0.526	0.298
85%	0.999	0.973	0.829	0.622	0.385	0.999	0.973	0.829	0.622	0.385
87%	1.000	0.987	0.889	0.719	0.485	1.000	0.987	0.889	0.719	0.485
89%	1.000	0.995	0.936	0.809	0.596	1.000	0.995	0.936	0.809	0.596
91%	1.000	0.998	0.969	0.886	0.712	1.000	0.998	0.969	0.886	0.712
93%	1.000	1.000	0.988	0.944	0.823	1.000	1.000	0.988	0.944	0.823
B. N,K	2,3	2,4	2,5	2,6	2,7	3,3	3,4	3,5	3,6	3,7
F	171875	245	0.4	0.0	0.0	708	0.0	0.0	0.0	0.0

(A) Columns are for N sizes of 2 and 3 and K sizes of 3–7. Rows represent various percentage identities between the homologous sequences. The table entries show the fraction of homologies detected. (B) N and K represents the number and size of the perfect matches, respectively. F shows how many perfect clustered matches expected to occur by chance in a translated genome of 3 billion bases using a query of 167 amino acids.

Selecting Initial Match Criteria

Both single imperfect matches and multiple perfect matches have a significant advantage over single perfect matches. They drastically reduce the number of alignments which must be checked to achieve a given level of sensitivity. The multiple-perfect match criteria can be modified to allow small insertions and deletions within the homologous area by allowing matches to be clumped if they are near each other rather than identical on the diagonal coordinate. This improves re-al-world sensitivity at the expense of increasing the number of alignments that must be done. Allowing a single insertion or deletion increases the alignments by a factor of three, whereas allowing two increases the alignments by a factor of five. In general, two perfect matches with the appropriate K size give specificity for a given level of sensitivity similar to that given by three or more perfect matches. The near-perfect match criterion overall is similar to the two perfect match criteria. The near-perfect criterion cannot accommodate insertions or deletions, but it has superi-or performance on finding small regions of homology. For finding coding exons in mouse/human alignments, whichever strategy is used, greater specificity is seen at the amino acid rather than the nucleotide level.

EST Alignment Choices

	K	F
1 Perfect	14	399
Near-Perfect	22	0.3
2 Perfect	11	0.1
3 Perfect	9	0.0

Maximum K sizes and number of chance matches passed to the alignment stage when searching for 100 bases of 95% homology with at least a 99% chance of detecting the homology. These values reflect our targets for EST alignments.

Mouse/Human Alignment Choices

	K	F	F Translated
1 Perfect-DNA	7	13,078,962	
1 Perfect-AA	5	62,625	187,875
Near-perfect-DNA	12	275,671	
Near-perfect-AA	8	749	2,247
2 Perfect DNA	6	237,983	
2 Perfect AA	4	245	734
3 Perfect DNA	5	109,707	
3 Perfect AA	3	708	2,123

Assuming 86% base identity and 89% amino acid identity, this table shows the maximum K sizes and number of chance matches passed to the alignment stage when searching for regions of 100 bases (or 33 amino acids) with at least a 99% chance of detecting the homology. These values reflect our targets for human/mouse alignments. For translated DNA sequences, the F value is multiplied by six to reflect three reading frames on both strands of the query. Even with this mul-tiplication, the specificity for a given sensitivity is several orders of magnitude greater in the amino acid rather than the nucleotide domain.

Homology Size (In Nucleotides) vs. Sensitivity for Four Search Criteria as Applied to Mouse/Human Comparisons Assuming 86% Nucleotide Identity and 89% Amino Acid Identity

	20	30	40	50	60	70	80	90	100	110	120	130
A	0.625	0.805	0.881	0.927	0.962	0.977	0.986	0.993	0.995	0.997	0.999	0.999
B	0.000	0.394	0.687	0.851	0.932	0.932	0.970	0.987	0.995	0.998	0.999	0.999
C	0.483	0.733	0.862	0.929	0.963	0.965	0.981	0.990	0.995	0.997	0.999	0.999
D	0.000	0.783	0.783	0.953	0.953	0.953	0.990	0.990	0.998	0.998	1.000	1.000

The median size of a human exon is 120 bases (International Human Genome Sequencing Consortium 2001). A, perfect 5 base amino acid match; B, two perfect 4-base amino acid matches within 100 amino acids and on diagonal; C, near-perfect 12 nucleotide match; D, near-perfect 8 amino acid match.

Since single-base insertions or deletions are relatively common artifacts of the sequencing process, nucleotide BLAT uses the two perfect 11-mer match criteria by default. Table shows actual alignment times for nucleotide BLAT on a collection of ESTs at various settings. For protein matches, the default criterion is a single perfect 5 for the stand-alone program. This is because the extension phase of protein BLAT is extremely quick in the stand-alone program, so the false positives generated by this approach have relatively little cost. The client/server protein BLAT uses three perfect 4-mers by default because in the client/server version, a portion of the genome must be loaded from disk for each false positive, a relatively time-consuming operation. As a result, the client/server protein BLAT is somewhat less sensitive than the stand-alone version.

Alignment Times in Seconds of 10,000 ESTs (Average Size 380 Bases) Against Human Genomic Sequence Using Various K Sizes and N Sizes

K	N	2×10^6	2×10^7	2×10^8
10	2	3.9	35.6	680.1
10	3	3.2	21.4	348.7
11	2	2.4	8.1	92.4
11	3	2.3	6.5	61.8
12	2	3.9	7.0	39.9
12	3	3.7	6.4	33.8

The 2×10^6 genomic sequence is ctg12414, which is 2,034,363 bases long and was taken from the December 2000 UCSC human genome assembly. The 2×10^7 genomic sequence is ctg15424 and is 20,341,418 bases long. The 2×10^8 column is chromosome 4 and is 200,175,155 bases long. The two major components of the run-time are the time it takes to bin and sort the K-mer hits (clumping is almost instantaneous after sorting), and the time it takes to extend the clumps into alignments. The bin/sort time depends on the number of hits, which is proportional to 4^{-K}. The bin/sort time is somewhere between O(n) and O(n log n). The extend time is linear with respect to the number of clumps.

Clumping Hits and Identifying Homologous Regions

To implement the match criteria, BLAT builds up an index of nonoverlapping K-mers and their positions in the database. BLAT excludes K-mers that occur too often from the index, as well as K-mers containing ambiguity codes and optionally K-mers that are in lowercase rather than uppercase. BLAT then looks up each overlapping K-mer of the query sequence in the index. In this way, BLAT builds a list of "hits" where the query and the target match. Each hit contains a database position and a query position. The following algorithm is used to efficiently clump together multiple hits. The hit list is split into buckets of 64k each, based on the database position. Each bucket is sorted on the diagonal (database minus query positions). Hits that are within the gap limit are bundled together into proto-clumps. Hits within proto-clumps are then sorted along the database coordinate and put into real clumps if they are within the window limit on the database coordinate. To avoid missed clumps near the 64k bucket boundary, unclumped hits and clumps that are within the window limit are tossed into the next bucket for additional clumping opportunities. The sorting algorithm mSort, which is related to qSort, is used. The bucketing tends to keep N relatively small.

Clumps with less than the minimum number of hits are discarded, and the rest are used to define regions of the database which are homologous to the query sequence. Clumps which are within 300 bases or 100 amino acids in the database are merged together. Five hundred additional bases are added on each side to form the final homologous region.

Searching for Near Perfect Matches

BLAT has an option to allow one mismatch in a hit. This is implemented by scanning the index repeatedly for each K-mer in the query. Every possible K-mer that matches in all but one position, as well as the K-mer that matches at every position, is looked up. In all, $K*(A-1)+1$ lookups are required. For an amino-acid search with $K=8$, this amounts to 153 lookups. Because a straight index of 8-mers would require 20^8 index positions or about 100 billion bytes, it is necessary to switch to a hashing scheme rather than an indexing scheme, further cutting efficiency. As a consequence, for a given level of sensitivity, the near-perfect match criterion runs 15× more slowly than the multiple-perfect match criterion in BLAT. The near-perfect match criterion seems best suited for programs that hash the query sequence rather than the database. A query sequence is sufficiently small that each possible nearly matching K-mer could be hashed, and therefore the index would not have to be scanned repeatedly.

The Relative Sensitivity and Specificity of BLAT at Various Settings With and Without a Best-Match Filter

All nonhuman mRNA alignments		Near perfect	CPU days	% genome	% RefSeq	Enrichment
K	N					
5	1	no	8.69	0.90%	69%	77x
4	2	no	6.06	0.83%	68%	81x
8	1	yes	95.25	0.86%	69%	80x
Only the best alignment for each mRNA		Near perfect	CPU days	% Genome	% RefSeq	Enrichment
K	N					
5	1	no	8.69	0.70%	67%	95x
4	2	no	6.06	0.66%	65%	99x
8	I	yes	95.25	0.68%	66%	97x

The human genome (August 2001 freeze UCSC assembly) was aligned against a collection of 86,000 nonhuman mRNA sequences totaling 123,000,000 bases taken from Genbank. The CPU days were measured on 800 MhZ Pentiums. The percentage genome column shows what percent of bases in the human genome are part of gapless alignments with the mRNAs. The percentage RefSeq shows the percentage of coding bases in human RefSeq-defined genes (from the database at) which are part of a gapless alignment. Enrichment is the ratio of the RefSeq coding density inside of the alignments compared to coding density in the genome as a whole. Higher levels of enrichment indicate greater specificity. Also shown are the same statistics when only the single best place that a nonhuman mRNA aligns was considered.

Alignment Stage

The alignment stage performs a detailed alignment between the query sequence and the homologous regions. For historical reasons, the alignment stage for nucleotide and protein alignments is quite different. Both have limitations, and are good candidates for future BLAT upgrades. On the other hand, both are quite useful in their present form for sequences which are not too divergent.

Nucleotide Alignments

The nucleotide alignment stage is based on a cDNA alignment program first used in the Intronerator. The algorithm starts by generating a hit list between the query and the homologous region of the database. Because the homologous region is much smaller than the database as a whole, the algorithm looks for relatively small, perfect hits. If a K-mer in the query matches multiple K-mers in the region of homology, the K-mer is extended by one repeatedly until the match is unique or the K-mer exceeds a certain size. The hits are then extended as far as possible allowing no mismatches, and overlapping hits are merged. The extended hits that follow each other in both query and database coordinates are then linked together into an alignment. If there are gaps in the alignment on both the query and database side, the algorithm recurses to fill in these gaps. Because the gaps are smaller than the original query and database sequences, a smaller k can be used in generating the hit list. This continues until either the recursion finds no additional hits, or the gap is five bases or less. At this point, extensions through Ns, extensions that allow one or two mismatches if followed by multiple matches, and finally extensions that allow one or two insertions or deletions (indels) followed by multiple matches are pursued. For mRNA alignments, it is often the case that there are several equivalent-scoring placements for a large gap in the query sequence. Generally such gaps correspond to an intron. Such gaps are slid around to find their best match to the GT/AG consensus sequence for intron ends.

The nucleotide alignment strategy works well for mRNA alignments and the type of alignments needed for genomic assembly. In these cases, the sequence identity is typically 95% or better. The strategy starts to break down when base identity is below 90%, and is therefore not suitable for most cross-species alignments.

Protein Alignments

The protein alignment strategy is simpler. The hits from the search stage are kept and extended into maximally scoring ungapped alignments (HSPs) using a score function where a match is worth 2 and a mismatch costs 1. A graph is built with HSPs as nodes. If HSP A starts before HSP B in both query and database coordinates, an edge is placed from A to B. The edge is weighted by the score of B minus a gap penalty based on the distance between A and B. In the case where A and B overlap, a "crossover" point is selected which maximizes the sum of the scores of A up to the crossover and B starting at the crossover, and the difference between the full scores and the scores just up to the crossover is subtracted from the edge score. A dynamic program then extracts the maximal-scoring alignment by traversing this graph. The HSPs in the maximal-scoring alignment are removed, and if any HSPs are left the dynamic program is run again.

The major limitation of this protein alignment strategy is that if there is an indel, part of the alignment will be lost unless the search stage manages to find both sides of the indel. For the translated mouse versus translated human genome job, which was the major motivation for protein BLAT, this limitation is not as serious as it would be when searching for more distant homologs. Indeed in the translated mouse/translated human case, this limit on indels is actually useful in some ways as it reduces the amount of pseudogenes which are found by BLAT more than it reduces the amount of genes found. Even so, in the future we hope to replace this simplistic extension phase with a banded (only small gaps allowed) Smith-Waterman algorithm.

Stitching and Filling In

It is often the case that the alignment of a gene is scattered across multiple homologous regions found in the search phase. These alignments are stitched together using a minor variation of the algorithm used to stitch together protein HSPs. For DNA alignments at this stage, the gap penalty is equal to a constant plus the log of the size of the gap. For mRNA/genomic alignments, if after stitching there are gaps left between aligning blocks in both the database and query sequence, the nucleotide alignment algorithm is called on the gap to attempt to fill it in. This gives BLAT a chance to find small internal exons that are further away than 500 bases from other exons, and which are too small to be found by the search stage.

Since the sort time is O(N logN), that is, proportional to N times log N, where N is the number of hits to be sorted, and the dynamic program time is $O(N^2)$ where N is the number of HSPs, an additional step is necessary to make BLAT efficient on longer query sequences. Untranslated nucleotide queries longer than 5000 bases and translated queries longer than 1500 bases are broken into subqueries that have approximately 250 bases of overlap. Each subquery is aligned as above, and the resulting alignments are stitched together.

References

- Hmm-baum-welch-derivation: people.eecs.berkeley.edu, Retrieved 18 April 2018

- Selection-Methods-for-Genetic-Algorithms-259461147: researchgate.net, Retrieved 28 May 2018

- Genetic-algorithms-mutation: tutorialspoint.com, Retrieved 13 June 2018

- How-to-define-a-fitness-function-in-a-genetic-algorithm-be572: towardsdatascience.com, Retrieved 13 April 2018

- Basic-local-alignment-search-tool-blast-29096: nature.com, Retrieved 23 May 2018

## 5

Software used in Bioinformatics

The software used in bioinformatics range in complexity from simple command-line tools to complex graphical programs. This chapter explores the different software used in bioinformatics such as FASTA, Clustal, GeneMark, GenoCAD, RAPTOR, sequence profiling tool, bowtie, etc.

Clustal

Pair wise sequence alignment has been approached with dynamic programming between nucleotide or amino acid sequences. The same approach can be used for alignment of 'n' number of sequences. But this program is limited to pair wise, since there will be exponential increase in memory, number of steps with respect to number of sequences. Because of such limitations with dynamic programming, researchers came up with an approach called *'progressive method'* to align three or more sequences.

Progressive method was first suggested by Feng and Doolittle in 1987. It compares only a pair of sequences together at a time using the following steps:

1. Using the standard dynamic programming algorithm on each pair, we can calculate the $(N*(N-1))/2$ (N is total number of sequences) distances between the sequence pairs.

2. From the distance matrix obtained using the clustering algorithm, construct a guide tree.

3. From the tree obtained, align the first node to the second node. After fixing the alignment, add another sequence or the third node. Iterate the step until all the sequences are aligned. When a sequence is aligned to a group or when there is alignment in between the two groups of sequences, the alignment is performed that had the highest alignment score. The gap symbols in the alignment replaced with a neutral character. Where it helps to guide the alignment of sequence- alignment and alignment –alignment.

Working of Algorithm

Multiple sequence alignment can be done through different tools. CLUSTALW is one among the mostly accepted tool.

Higgins D has written the first program of CLUSTAL, considering memory and time various CLUSTAL series of programs have came up and presently used version is CLUSTALW, which came up with dynamic programming and progressive alignment methods.

CLUSTALW uses the progressive algorithm, by adding the sequence one by one until all the sequences are completely aligned.

Steps for CLUSTAL algorithm

1. Calculate all possible pairwise alignments, record the score for each pair.

2. Calculate a guide tree based on the pairwise distances (algorithm: Neighbor Joining).

3. Find the two most closely related sequences

4. Align the sequences by progressive method

    i. Calculate a consensus of this alignment

    ii. Replace the two sequences with the consensus

    iii. Find the two next-most closely related sequences (one of these could be a previously determined consensus sequence).

    iv. Iterate until all sequences have been aligned

5. Expand the consensus sequences with the (gapped) original sequences

6. Report the multiple sequence alignment

GenoCAD

In order to fully reap the potential benefits of *de novo* chemical gene synthesis it has become necessary to develop tools and methodologies to streamline the design of custom DNA sequences. Protein expression for structural studies functional genomics, metabolic engineering, or gene expression studies are only some of the numerous possible applications of this emerging technology. Beyond small scale genetic constructs encompassing no more than a few interacting genes, it becomes possible to reengineer viral, bacterial, and even eukaryotic genomes. While the number of users of this technology increases, so does the need to streamline the design of synthetic DNA sequences. GenoCAD is a web-based application filling this need by providing users with an integrated graphical development environment that no other software provides.

GenoCAD's design philosophy derives from the notion of genetic parts, which was first articulated to analyze genomics data. Thinking of genetic systems as composed of parts, each with its own function and characteristics is akin to the way parts are described and used in various engineering fields. Designing complex systems through a bottom up integration of components is a dominant paradigm in engineering. It was therefore natural that engineers approaching DNA as an engineering substrate, rather than a natural macromolecule, used the notion of biological parts as building blocks. For instance, promoters, ribosome-binding sites (RBS), genes and terminators are all categories of parts that are needed for designing complex prokaryotic genetic constructs such as switches and oscillators . One could argue that systematic efforts to decompose biological sequences into functional modules that can be recombined to meet user-defined specifications is one of the most distinctive features of synthetic biology compared to more traditional uses of recombinant DNA technologies.

GenoCAD facilitates the design of artificial DNA sequences in three ways. First, GenoCAD includes a flexible system to manage libraries of public and user-defined genetic parts. Second, GenoCAD relies on formal design strategies to guide both novice and experienced users in the design of structurally valid constructs for various biological applications. Finally, GenoCAD's sophisticated data model enables individual users and research groups to customize their workspace to their specific needs.

Flexible Management of Genetic Parts Libraries

Nothing better attests the benefits of a parts-based approach to the design of genetic constructs than the success of the Registry of Standard Biological Parts. By defining the BioBrick™ standard allowing the composition of parts and implementing mechanisms to share parts, the Registry has been critical in fostering the development of a vibrant synthetic biology community. We recently analyzed the content of the Registry database and the associated collection of clones to better understand how the successes and limitations of this pioneering experiment could guide the development of a second generation of registries of biological parts. GenoCAD attempts to refine some of the concepts upon which the Registry was developed.

By assuming that genetic designs can be synthesized, GenoCAD eliminates the need for standardizing the means by which parts are connected. It also eliminates the need to develop a collection of bacterial clones to manage the physical implementation of the parts. Our analysis also stressed the importance of basic parts used to generate new combinations of parts with specific functions. Ensuring the accuracy of the sequence and annotation of the basic parts is essential since inaccuracies at this level may affect numerous designs. As a result of this observation, GenoCAD parts libraries are exclusively composed of basic parts while sequences composed of multiple parts are called designs. The libraries of parts available to all GenoCAD users are limited to sequences used in peer-reviewed publications or commercial vectors. Parts are curated by a small number of experts.

Categorizing parts into functional groups has also proved challenging as the number and diversity of parts increases. It would be, for instance, questionable to record bacterial and eukaryotic promoters in the same group. Developing a more granular categorization system may lead to an exponential growth of categories that would prove cumbersome to navigate. GenoCAD overcomes this challenge by relying on the notion of grammar. A grammar is composed of rules describing the structure of DNA sequences. One of the rules of the *Escherichia coli* Expression Grammar is CAS→PRO CIS TER which reads: an expression cassette is composed of a promoter, a cistron and a terminator. Another rule of the same grammar is CIS→RBS GEN (a cistron is composed of a RBS and a gene). The two rules can be used successively to create a basic expression cassette PRO RBS GEN TER. Different grammars can be developed for different applications and each grammar has its own parts categorization hierarchy. This approach ensures, for instance, that parts suitable for designing constructs for specific organisms like *E. coli* or yeast can easily be identified. It also enables the development of grammars for specific applications like protein production, homologous recombination in yeast, etc. Instead of attempting to develop a universal parts categorization system, GenoCAD provides a generic framework for the development of smaller more manageable application-specific parts-libraries. The 'Parts' tab of the GenoCAD user interface provides a parts library browser.

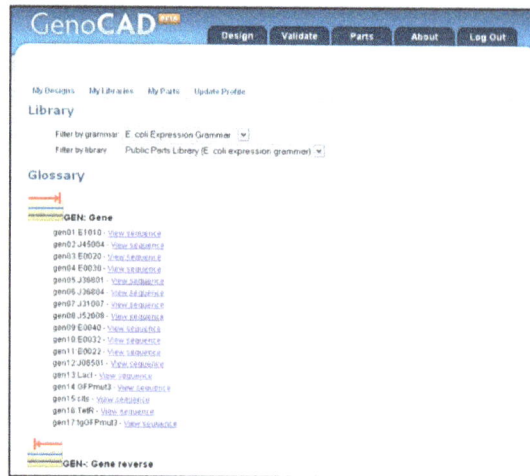

The GenoCAD parts library browser. Parts are associated with individual libraries, each of which is associated with a specific grammar. Users select which parts library they view through choice of a grammar and specific library in drop down boxes on the page. The part category 'Gene' is displayed in this figure along with the icon that represents genes in the designs. By clicking on the icon, the list of genes expands, allowing the user to see the available choices in the library. Selecting the link to 'View Sequence' for any part opens a small window containing the sequence of the individual part.

Point-and-Click Design of Genetic Constructs

In addition to providing a hierarchy of categories, grammars include sets of rewriting rules that formalize design strategies for various types of genetic constructs. The design feature of GenoCAD embeds the grammars in a graphical user interface that leads users through the design workflow formalized in the grammar. Grammars usually prompt users to begin by choosing high-level structures of their system and systematically decomposing them into individual part categories. The last step of the design process consists in selecting actual parts corresponding to specific DNA sequences. When starting from one of the public design templates, users can quickly design constructs by simply selecting parts in a parts library instead of going through the entire design process described below.

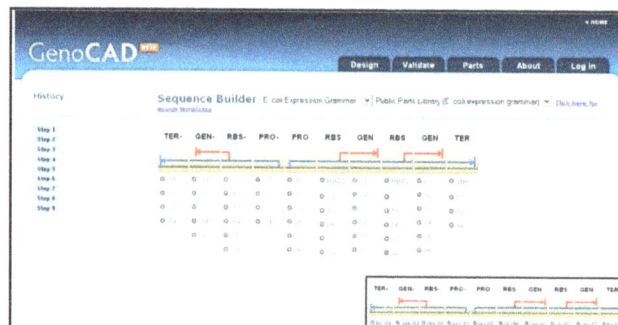

The design interface showing the structure of a genetic toggle switch. The interface has drop down boxes at the top to select the grammar and parts library that will be used in the design. The history panel allows users to select one of the steps in the design process and see the structure of the design at that step. Users are permitted to go back to any step and redesign from that point. The design is presented in the main panel of the page, and icons for each part and the abbreviated parts categories are shown at the top of the design. Choices for each part are shown underneath the part icon. The inset shows the final design for this construct after specific choices (terminals) are selected for each part category.

Here, we use the design of a bistable genetic switch to illustrate GenoCAD's workflow. Selection of a grammar and an associated parts library is the first step of this process. By selecting the grammar the user defines the type of construct that is possible, and by selecting the library they define the set of parts available to complete the design. Determining the high level structure of a functional genetic system *de novo* could potentially be confusing to users that do not have an intimate knowledge of the role of each part in the regulation of gene expression. However, each grammar behind GenoCAD provides the design strategy for specific types of constructs. GenoCAD's default grammar, 'E. *coli*-Expression Grammar', is suitable for the design of prokaryotic expression constructs. Additional grammars will be added for other applications.

The toggle switch construct was designed in nine steps as shown in. The history pane (left side) allows users to review their work at any stage of the design process by clicking on that step. Users may click in reverse or in forward steps, and they are able to redesign from any point if they wish. Shows a slightly condensed version of the design process of the toggle switch construct. Displayed below each part of a design are the options available to the user to transform the design. Choices in gray correspond to transformations that affect the design structure while options in white correspond to the selection of a specific part. Structural transformations are either part categories for which a specific part selection can be made or a higher level feature that must be decomposed to features lower in the abstraction hierarchy before the design can be finalized. An example of a high level part is the cistron, which is transformed into a RBS and gene (GEN). For each part category users have a choice of one or more specific sequences such as a specific RBS or gene. A mouse-over feature in the interface provides more information about the available choices. For example, the name of choice 04 for the promoter (PRO) category is displayed through this mechanism.

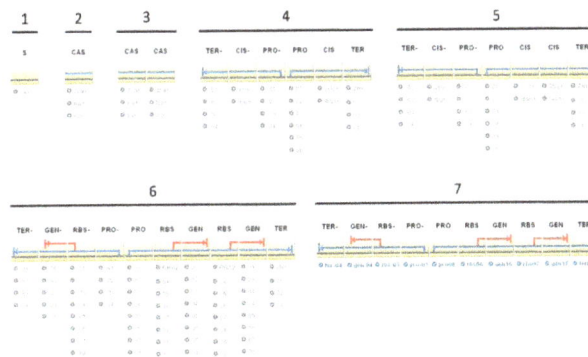

A progression through the design of a bistable toggle switch. The starting symbol, S, is where each design begins, and it is transformed into a transcription cassette, CAS,. Since the toggle switch contains two transcription cassettes, the single cassette is doubled. The design we are following has the cassettes oriented in opposite directions, and we achieve this by transforming the left cassette to the tpc- option, and the right to the tcp+ option, which contain a promoter, cistron, and terminator, but in opposite orientations. The right cassette is meant to express a transcriptional repressor and reporter gene in a bicistronic manner, so the cistron is doubled by selecting the 2cis+ option. Each cistron is then decomposed to a RBS and gene, with the RBS and gene in the reverse orientation in the left cassette. Selection of the specific promoters, RBSs, genes, and terminators produces a final construct that is associated with a DNA sequence.

A design is complete when specific sequences have been selected for each of its structural features. It is then possible to click on the 'Download Sequence' button to export the construct sequence as a text file that can be imported into software to design oligos for gene synthesis. Alternatively the sequence synthesis can be ordered from a fast growing number of vendors providing contract gene synthesis services.

Implementation and Data Model

The GenoCAD website is written in a combination of PHP and JavaScript and runs on an Apache server. The MySQL database is on a different server, and both servers use the Linux operating system. The validation page relies on a custom parser developed in C++.

The data model for GenoCAD is summarized in the figure below. Each design is associated with a specific parts library which, in turn, is linked to a specific grammar. Multiple public and user-defined libraries can be associated with each grammar and multiple designs can be associated with a specific library. Parts defined by users need to be associated with one of the user's parts libraries.

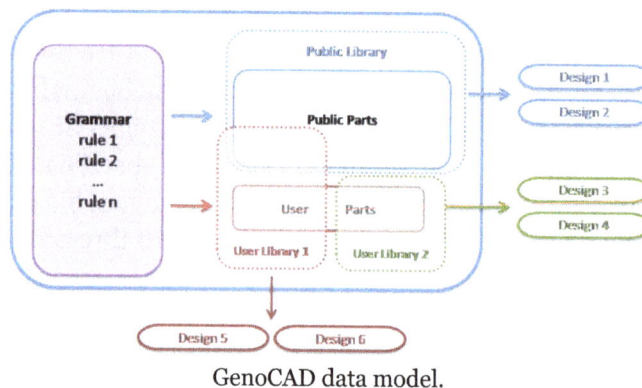

GenoCAD data model.

Each grammar encompasses a set of rules by which constructs can be designed. The grammar also defines the categories of parts that are available to design the constructs. For each grammar there is a collection of public parts (solid, blue rectangle), which constitute a publicly available parts library (dashed, blue rectangle). 'User Libraries' can be created from any subset of the public parts, and this library can be supplemented with user-created parts (solid, red rectangle). Two user libraries (dashed, red and dashed, green rectangles) are shown here that contain different subsets of public and user-created parts. User library 2 contains all user-created parts. When a design is created, all the parts to complete the design must be contained within a single library.

This simple data model has several limitations. Since many parts such as coding sequences can be used in different organisms, it would be desirable to replace the current hierarchical data model with a more refined model allowing the same part to be used in multiple libraries and grammars. It would also be desirable to define parts corresponding to coding regions by their amino-acid sequences instead of being limited to a DNA sequence with codons that are optimal for expression in a specific organism.

Additional grammars will be defined for the use of specific applications or for applications relevant to specific organisms in collaboration with organism or domain experts. Defining parts categories and design strategies suitable for a particular application will require a dialogue between biologists and computer scientists having experience in grammar development. Once agreed upon, grammars can easily be recorded in the MySQL database.

GeneMark

Both GeneMark and GeneMark.hmm can be used via the GeneMark website for the analysis of prokaryotic DNA, with 175 pre-computed species-specific statistical models available. Analysis of DNA from any prokaryotic species is supported by (i) a special version of GeneMark.hmm using a heuristic model calculated from the nucleotide frequencies of an input sequence at least 400 nt long and (ii) a self-training program, GeneMarkS, which can be used for longer sequences on the order of 1 Mb in length. Thus, the DNA of any prokaryote can be analysed, via either a pre-computed species-specific model or a model created on the fly.

As many of the programs at the GeneMark website share similar interfaces, we use here the prokaryotic GeneMark.hmm program as an exemplar and discuss program-specific differences below, where appropriate.

The GeneMark.hmm web interface accepts as input a single DNA sequence as an uploaded file or as text pasted into a textbox. If a FASTA description line begins the sequence, all text on the line following the 'greater than' symbol (>) is used as the title. In the remainder of the submission, digits and white space characters are ignored and letters other than T, C, A and G (assumed to appear rarely) are converted to N. The interface requires selection of the species name. Selection of a model for the RBS (in the form of a position-specific weight matrix and a spacer length distribution) is optional. In certain cases, such as the crenarchaeote *Pyrobaculum aerophilum*, the RBS model is replaced by a promoter model, which is the dominant regulatory motif located upstream to gene starts in this species. The interface also includes the option of using other types of genetic codes such as the Mycoplasma genetic code.

GeneMark.hmm reports all predicted genes in a format that includes the strand the gene resides on, its boundaries, length in nucleotides and gene class. Class indicates which of the two Markov chain models used in GeneMark.hmm, Typical or Atypical gene model provided the higher likelihood for the gene sequence. Genes of the typical class exhibit codon usage patterns specific to the majority of genes in the given species, while atypical class genes may not follow such patterns and frequently contain significant numbers of laterally transferred genes. The nucleotide sequences of predicted genes and translated protein sequences are available as an output to facilitate further analysis, such as BLAST searching. An option to generate GeneMark predictions in parallel with the GeneMark.hmm analysis provides important additional information. In this case, GeneMark is set up to use models derived from the same training data as models for the current run of GeneMark.hmm.

It is worth noting that the GeneMark.hmm and GeneMark algorithms are complementary to each other in the same way as the Viterbi algorithm and the posterior decoding algorithm are. Therefore, though the two algorithms are distinct, they are supposed to generate predictions largely corroborating and validating each other. Differences frequently indicate sequence errors and deviations in gene organization, very short genes, gene fragments, gene overlaps, etc.

Graphical output of the analysis is available in PDF or PostScript format. A fragment of this output, illustrating the predictions of both GeneMark and GeneMark.hmm, is shown in Figure. The graphical output clearly depicts the advantage of using multiple Markov chain models representing different classes of genes. Here, the coding potential graph obtained using the Typical gene model, derived by GeneMarkS, is denoted by a solid black line, and the coding potential graph obtained using the Atypical gene model (derived by a heuristic approach) is denoted by a dotted line. Whereas the first and last genes could be detected using either of the two models, as both of them produced high enough coding potentials, the gene located in positions from 846 to 1112 was detected only by the atypical model. Further, figure demonstrates the ability of the GeneMark programs to detect genes of both the Typical and Atypical gene classes. The GeneMark graph also includes indications of frameshift positions, which are often sequencing errors but in rare cases are natural and biologically very interesting.

For the GeneMark program, there are several specific options. The window size and step size parameters (96 nt and 12 nt, respectively, by default) define the size of the sliding window and how far this window is moved along the sequence in one step. The threshold parameter determines the minimal average coding potential for an open reading frame (ORF) to be predicted as a gene. There are several options which allow fine-tuning of the GeneMark graphical output. In addition, there are options supporting the analysis of eukaryotic DNA sequences by GeneMark including the ability to provide lists of putative splice sites and protein translations of predicted exons. As might be expected, GeneMark (the posterior decoding algorithm) does not produce high enough resolution for the precise prediction of exon–intron borders. Thus, GeneMark.hmm (the generalized Viterbi algorithm) in its eukaryotic version is the major tool for the identification of exon–intron structures in eukaryotic DNA sequences.

The output of the GeneMark program consists of a list of ORFs predicted as genes, i.e. those with average coding potential above the selected threshold. Although each predicted gene can have more than one potential start, additional data is provided to help the researcher annotate one of the alternatives as the 'true' one. The start probability (abbreviated 'Start Prob') is derived from the sequences in the windows immediately upstream and downstream of each potential start. RBS information is provided in the form of a probability score along with the position and sequence of the potential RBS (abbreviated 'RBS Prob', 'RBS Site' and 'RBS Seq'). In addition to the list of predicted genes, GeneMark provides a list of 'regions of interest', spans of significant length between in-frame stop codons where spikes of coding potential are wide enough and may warrant further analysis even if no genes are predicted therein based on automatic comparison with the threshold.

Analysis of prokaryotic DNA sequences for which there is no pre-computed species-specific model can be carried out using a program version which heuristically derives a model for any input sequence >400 nt. This approach has also proven useful for the analysis of inhomogeneous genomes, particularly regions too divergent from the bulk of the genome, such as pathogenicity islands.

If models (including RBS models) have to be computed *de novo* for an anonymous DNA sequence with length of the order of 1 Mb or longer, the GeneMarkS program can be used. This program needs significantly more computational resources; thus, its output is provided via email. A modified version of GeneMarkS tuned for the analysis of viruses of eukaryotic hosts creates a model for the Kozak consensus sequence instead of a two-component RBS model.

Gene predictions made by the prokaryotic version of GeneMark.hmm for a fragment of the *Escherichia coli* K12 genome

Gene	Strand	Left end	Right end	Gene length	Class
1	+	61	825	765	1
2	+	846	1112	267	2
3	+	1145	2092	948	1
4	–	2254	4386	2133	1
5	–	4388	520	366	1

In the 'Class' column, 1 and 2 indicate Typical and Atypical, respectively. Direct and reverse complement strands are indicated by '+' and '–', respectively. The graphical output for the first three predictions is shown in Figure.

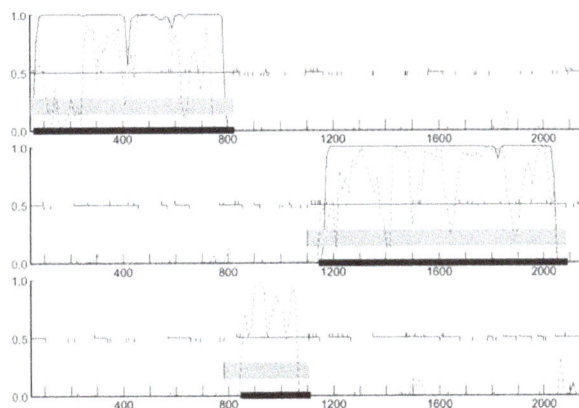

Graphical output from the combination of GeneMark and GeneMark.hmm for a fragment of the *Escherichia coli* K12 genome. The solid black and dashed traces indicate the coding potential calculated by the GeneMark program using the Typical and Atypical Markov chain models of coding DNA, respectively. Only the three reading frames in the direct strand are shown as there are no genes (either predicted or annotated) on the reverse strand in this section of the genome. The thick black horizontal bars indicate the locations of the predictions made by GeneMark.hmm. The thick grey horizontal bars indicate 'regions of interest' provided by the GeneMark program. The thin black horizontal lines indicate (longest) ORFs observed in each reading frame; ticks extending above and below this line indicate potential start and stop codons, respectively.

SEQUEST

As with most bioinformatics algorithms, SEQUEST had its origins in a cumbersome manual process. A seminal paper from Don Hunt in 1986 illustrated the challenges of interpreting peptide tandem mass spectra. John Yates, then a graduate student in the Hunt laboratory, began thinking of ways to apply computers in the process of spectral interpretation and built upon that experience during his early years as a faculty member. Kevin Owens' 1992 review of correlation analysis in mass spectra provided a mechanism by which tandem mass spectra could be compared to each other, and John Yates hired Jimmy Eng, an electrical engineer who had recently completed his Master's degree at the University of Washington, to begin software development in earnest.

SEQUEST was effective because of a series of shrewd judgment calls in software development. Sequence databases were miniscule, by today's standards (the *S. cerevisiae* genome was not completed until 1996). Dr. Yates, however, recognized early that using predicted protein sequences from genomic sequencing would drastically reduce the set of potential sequences to be compared to each tandem mass spectrum. Similarly, the group recognized that predicting the appearance of collision-induced dissociation (CID) tandem mass spectra accurately for peptide sequences was a daunting challenge, and they opted to employ very simple fragmentation models that predicted C-terminal y ions to be twice the intensity of N-terminal b ions. Each experimental spectrum was separated into ten zones by m/z, with peak intensities normalized within each to make the experimental spectra look more like the theoretical ones. Finally, they recognized that cross-correlation required so much CPU power that a pre-scoring routine was necessary to retain only 500 candidate peptides for full scoring by cross-correlation. Taken together, these insights paved the way for fully automated peptide identification software.

Making SEQUEST widely available led through gates of intellectual property, commercialization, and publication. On March 14, 1994, the University of Washington filed for a pair of patents (US5538897A and US6017693A) that defined the use of database searching for amino acid and nucleotide sequences from tandem mass spectra collected in mixtures of proteins. In 1993, Dr. Yates had begun discussions with Adrian Land and Ian Jardine, researchers at Thermo Instrument Systems (now Thermo Fisher Scientific), to commercially distribute the SEQUEST software. The University of Washington agreed to an exclusive license of the patents to Thermo Instrument Systems. Jim Shofstahl integrated the software into the DECUnix-based BioWorks for the TSQ 700 under the name "PepSearch" (the name "SEQUEST" was coined after the 1994 publication). At first this appeared to be an ideal solution, but later implementations separated SEQUEST from BioWorks so that updates to the rapidly-changing SEQUEST could be incorporated more readily.

Publishing SEQUEST, however, proved to be a significant challenge. Initially, the manuscript was sent to the *Proceedings of the National Academy of Sciences*, but the reviewers found it to be a mismatch for the journal. Dr. Yates then turned to *Protein Science*, which consulted the same reviewers as for *PNAS* in order to speed the process of review. The speedy review, however, resulted in rejections there, as well. Dr. Yates consulted with his mentor, Don Hunt, who advised publication in the *Journal of the ASMS* after consulting with Michael Gross. *JASMS* received the manuscript along with its prior reviews, and the paper was accepted only 27 days after its receipt on June 29, 1994. Later in the same year, Mann and Wilm published the manual interpretation sequence-tagging approach to peptide identification. That these two technologies were presented in the same year is no coincidence; tandem mass spectrometry was clearly the most promising data source for protein identification, and bioinformatics advances were critical to realizing its potential.

Interpreting SEQUEST results, of course, required additional tools. Thermo Instrument Systems had begun by licensing basic support tools, such as the "Display Ions" Peptide-Spectrum Match

(PSM) viewer and "SEQUEST Summary" result table builder, from the University of Washington. They soon licensed the Harvard Proteomics Browser Suite (licensed as the SEQUEST Browser), a growing collection of scripts from the William S. Lane Laboratory. These tools provided essential capabilities for the interpretation of data sets, such as the depth of protein sequence coverage in the Protein Report, between-experiment comparisons in IonQuest, and the recognition of variant peptide forms in MuQuest. The software assisted the manual interpretation of tandem mass spectra through the FuzzyIons tool and combined SEQUEST scores for better discrimination in the ScoreFinal neural network. In several respects, the Suite prefigured later identification workflows such as the Trans-Proteomic Pipeline. With these tools in place, the stage was set for large numbers of researchers to benefit from database searching.

Evolving New SEQUEST Capabilities and Applications

For the next seven years, the Yates Lab worked closely with Thermo to update SEQUEST continuously with improvements. The most essential boost came from the addition of "dynamic modifications". The software could be notified that certain amino acids may sometimes carry additional mass due to a post-translational modification. The initial searches with this feature were limited to dynamic PTMs on only two residues at a time. Soon thereafter, the number of modifiable residues was increased to three. With dynamic PTMs, SEQUEST came of age.

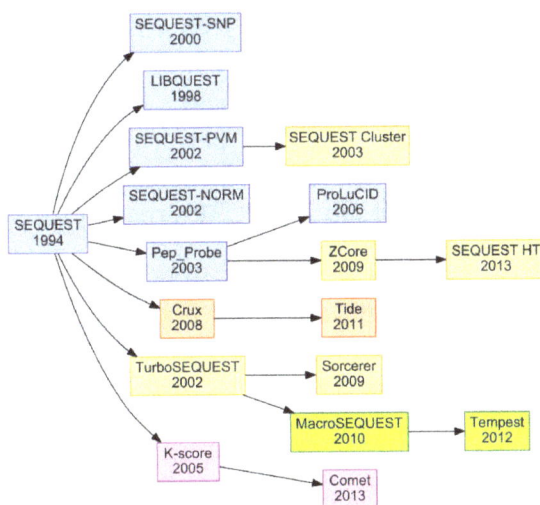

A tree representing the descendants of the original SEQUEST algorithm. Blue algorithms were produced in the Yates Laboratory, while yellow were produced in conjunction with commercial partners Thermo Fisher Scientific or Sage-N Research. Orange represents developments in the Noble Laboratory, with green denoting developments in the Gerber Laboratory and purple marking advances from Jimmy Eng after the year 2000. An arrow does not imply direct use of source code.

Early efforts in proteogenomics were also demonstrated in 1995 with the new ability to search nucleotide databases through six-frame translation. Dr. Yates was able to demonstrate that protein identification was feasible using the chromosome II, III, and IX sequences produced by the in-progress *S. cerevisiae* genome project. This paper also supplied an early answer for how relative scoring can be used to determine which spectra had been successfully identified; the paper specified that PSMs in which the best match scored 10% better than the second (a ΔC_n or DeltaCN greater than 0.1) could be trusted. Leveraging genomic data would later be augmented in SE-QUEST-SNP, which introduced non-synonymous single nucleotide polymorphisms to nucleotide databases for recognizing amino acid variants.

SEQUEST had become increasingly associated with the Thermo LCQ 3D ion trap after its release in 1996. The software was included in the new "XCalibur" interface for Windows NT. In an effort to make the software more broadly applicable, Yates Lab added the capability to look for a broader set of fragment ions associated with high-energy CID; similarly, they examined post-source decay spectra as a source of identifications. The Yates team also turned their efforts toward adapting the technique for spectral library searching. Their introduction of LIBQUEST applied the same PSM scoring system from SEQUEST to the matching of previously identified spectra with recently collected MS/MS scans.

Algorithm Efficiency and Parallelization

Improving the speed of SEQUEST execution was a priority from early in development. Modifications to the initial C++ codebase targeted both Windows and UNIX platforms. Cross-correlation is a powerful match discriminator, but it requires a computationally expensive Fast Fourier Transform (FFT) operation. The initial implementation was based on code from *Numerical Recipes in C* and from *Dr. Dobb's Journal*. At the time, floating-point performance for Intel processors was relatively slow, and so 64-bit DEC Alpha processors were investigated to improve execution. Over the next several years, though, Intel and AMD greatly improved performance by switching to 64-bit datapaths, accelerating math by operating on vectors of numbers, and increasing processor frequencies. These shifts have benefited SEQUEST performance even as MS/MS data sets have dramatically grown in size; during the time required for MS/MS scan rates to quintuple from an LCQ to an LTQ, the number of transistors in Intel CPUs rose by an order of magnitude from the 200 MHz Pentium Pro to the 2.4 GHz Pentium 4.

The pressures to improve search speeds continued, however, and Yates Lab and Thermo worked together and separately to address the problem. Jimmy Eng and Bill Lane each worked on strategies for pre-indexing FASTA sequence databases to sort the masses of tryptic peptides prior to search. Jim Shofstahl adapted this code to produce indexes which could be exploited for PTM searches, releasing "TurboSEQUEST" in BioWorks 3.0. Jimmy Eng was able to leverage the Parallel Virtual Machine package from ORNL to distribute the identification task across multiple computers, bridging between Windows master nodes and UNIX slave nodes. The SEQUEST-PVM software was able to accelerate these searches by a factor that scaled linearly with the number of computers in the cluster. Jim Shofstahl at Thermo tuned this software for more robust operation, and end users were able to purchase SEQUEST Cluster licenses with BioWorks 3.1 in which IBM provided computers and Thermo contributed necessary software. Under license with Thermo, Sage-N Research produced "Sorcerer," an FPGA (field-programmable gate array) that had been configured to accelerate FFT in hardware. Over time, Sage-N switched to source code optimizations in TurboSEQUEST to improve performance in x86 systems provided by the company.

Thermo and the University of Washington have occasionally licensed the TurboSEQUEST source code to universities. Vanderbilt University, for example, compiled the source to produce specialized executables; the campus supercomputing facility employed IBM JS20 blades that used PowerPC 970 processors rather than x86 or Alpha CPUs. The Gerber Lab at Dartmouth, however, had more ambitious ideas for their collaboration. Under their source license, the group produced MacroSEQUEST, a streamlined build of the software that searched all spectra simultaneously rather than using the spectrum-at-a-time approach of the original SEQUEST. A key modification made in MacroSEQUEST allowed for users to adjust the FFT bin size, which permitted users of HCD (a collision cell fragmentation that is similar to beam-type CID) high-resolution tandem mass spectra to profit from high fragment mass accuracy in XCorr computation. The group continued their modifications in the Tempest project to off-load cross-correlation to a graphical processing unit (GPU) or employ extremely fast dot product computation for scoring instead.

Thermo, of course, has continued to invest in development. SEQUEST-HT, which became available as part of Proteome Discoverer 1.4, is a reimplementation of the TurboSEQUEST algorithm using the Microsoft .NET framework. SEQUEST-HT is multi-threaded to take advantage of multi-core CPUs, now commonplace. It benefits from sequence database management in the ZCore algorithm along with its handling of ETD and HCD fragmentation with the small FFT bin sizes like those of MacroSEQUEST.

Search Engines from the Diaspora

After Yates Lab moved from the University of Washington to The Scripps Research Institute in the year 2000, intellectual property issues prevented the group from producing new variants of the software and publishing them as SEQUEST (which is a trademark owned by the University of Washington). Similarly, Jimmy Eng moved to the Institute for Systems Biology, shifted to the Fred Hutchinson Cancer Research Center in 2004, and then returned to the University of Washington in 2007.

At Scripps, the Yates Lab was increasingly encountering very large data sets as it employed fractionated sample techniques such as MudPIT. SEQUEST had been crafted to read individual MS/MS scans from DTA files and to write PSMs for individual spectra to OUT files, a strategy that led to significant file system problems when combining tens or hundreds of high-scan-rate LC-MS/MS experiments. Because of the bloat associated with XML file formats, the Yates Lab adopted delimited text formats for storing information: MS1 (mass spectra), MS2 (tandem mass spectra), and SQT (SEQUEST outputs) . Thinking along similar lines, Jim Shofstahl had created the binary SRF format for storing DTA and OUT data structures for the commercial SEQUEST release. Support for the HUPO-PSI mzML format was added to SEQUEST-HT via an importer in Proteome Discoverer, and output from SEQUEST-HT can be converted to mzIdentML format within that framework, as well.

Within Yates Lab, peptide identification included insights from Michael MacCoss, Rovshan Sadygov, and Tao Xu. In 2002, Michael MacCoss introduced SEQUEST-NORM, a variant of the SEQUEST code that could produce peptide length-independent cross-correlation scores. Dr. Sadygov incorporated the "Fastest Fourier Transform in the West" library into SEQUEST to accelerate FFT computation and added a dot product score for use with accurate mass MS/MS scans, with the enthusiastic support of Michael Senko at Thermo Fisher Scientific. Dr. Sadygov's experiments on improving the pre-scoring routines of SEQUEST led to an altogether new search engine; Pep_Probe employed a hypergeometric distribution rather than cross-correlation as its primary match score. This software was a useful test-bed for exploring other scoring functions. In 2005, Dr. Sadygov demonstrated the implementation in Pep_Probe of a scoring model based on accounting for larger fractions of total fragment ion intensity for an MS/MS, compared to cross-correlation and the original hypergeometric implementation. After Dr. Sadygov was employed by Thermo Fisher Scientific, he turned those skills to the identification of ETD spectra. In collaboration with the Coon Laboratory at the University of Wisconsin-Madison, he published the ZCore algorithm, which combined his hypergeometric assessment of matched peak counts with an assessment of the matched fragment ion intensities.

Tao Xu produced the current algorithm employed for database search in Yates Lab. ProLuCID employs the binomial distribution for determining the best 500 peptide sequences from a protein database and then applies cross-correlation to this set. The software predicts fragments with improved isotope models for better cross-correlation scoring discrimination. For each spectrum, ProLuCID determines the Z score for the highest XCorr against the distribution produced by the top 500 candidates, determining the extent to which the best match falls outside the distribution produced by random matches. In addition to dynamic modifications on particular residues, the software adds the capability for peptide N-terminal and C-terminal modifications. ProLuCID is written in Java and can be deployed on individual computers or on Linux clusters.

At the Institute for Systems Biology, Jimmy Eng began work on the Comet search engine in 2001. At first, it took the form of search engine that scored PSMs by inferring a Z-score from a distribution of dot-product scores instead of the costly cross-correlation of SEQUEST (much as Tempest did for HCD several years later) . The approach gained broader use as the "K-score" in X.Tandem . Upon his return to the University of Washington in 2007, Jimmy Eng returned to the SEQUEST scoring approach to discover a method by which FFT could be entirely bypassed in high-speed computation of cross-correlation scores a technique incorporated into SEQUEST-HT. Four years later, he had written the Comet search engine from the ground up to support standard file formats for inputs and outputs, support a variety of activation methods, and distribute processing over multiple threads. Comet reports expectation values that estimate how many PSMs might have been expected to score as well as the best match by random chance alone.

The University of Washington continued development in the SEQUEST family after the departure of Yates Laboratory, principally in the laboratory of William Noble. Christopher Park introduced Crux in 2008, featuring efficient peptide indexing for FASTA databases, on-the-fly decoy generation, and distribution fitting for top XCorrs. Crux paired efficiently with the Percolator algorithm from the Noble Lab for improved PSM discrimination. Benjamin Diament applied a wide variety of optimization strategies to create the highly efficient Tide algorithm, showing considerable improvements in search times compared to 1993 and 2009 builds of SEQUEST and to Crux. Further refinements in 2014 enabled the calculation of accurate p-values from XCorr scores. These tools were combined in the broader framework of the Crux Toolkit in 2014.

SAMtools

With the advent of novel sequencing technologies such as Illumina/Solexa, AB/SOLiD and Roche/454, a variety of new alignment tools have been designed to realize efficient read mapping against large reference sequences, including the human genome. These tools generate alignments in different formats, however, complicating downstream processing. A common alignment format that supports all sequence types and aligners creates a well-defined interface between alignment and downstream analyses, including variant detection, genotyping and assembly.

The Sequence Alignment/Map (SAM) format is designed to achieve this goal. It supports single- and paired-end reads and combining reads of different types, including color space reads from AB/SOLiD. It is designed to scale to alignment sets of 10^{11} or more base pairs, which is typical for the deep resequencing of one human individual.

Methods

SAM Format

The SAM format consists of one header section and one alignment section. The lines in the header section start with character '@', and lines in the alignment section do not. All lines are TAB delimited. An example is shown in Figure.

```
(a) coor   12345678901234   5678901234567890123456789012345
    ref    AGCATGTTAGATAA**GATAGCTGTGCTAGTAGGCAGTCAGCGCCAT

    r001+          TTAGATAAAGGATA*CTG
    r002+        aaaAGATAA*GGATA
    r003+      gcctaAGCTAA
    r004+                      ATAGCT.............TCAGC
    r003                          ttagctTAGGC
    r001                                    CAGCGCCAT

(b) @SQ SN:ref LN:45
    r001 163 ref  7 30 8M2I4M1D3M  = 37  39 TTAGATAAAGGATACTA *
    r002   0 ref  9 30 3S6M1P1I4M  *  0   0 AAAAGATAAGGATA    *
    r003   0 ref  9 30 5H6M        *  0   0 AGCTAA      *  NM:i:1
    r004   0 ref 16 30 6M14N5M     *  0   0 ATAGCTTCAGC       *
    r003  16 ref 29 30 6H5M        *  0   0 TAGGC      *  NM:i:0
    r001  83 ref 37 30 9M          = 7 -39 CAGCGCCAT         *

(c) ref  7 T 1 .     ref 12 T 3 ...    ref 17 T 3 ...
    ref  8 T 1 .     ref 13 A 3 ...    ref 18 A 3 . 1G..
    ref  9 A 3 ...   ref 14 A 2 .+2AG.+1G  ref 19 G 2 *.
    ref 10 G 3 ...   ref 15 G 2 ..    ref 20 C 2 ..
    ref 11 A 3 ..C   ref 16 A 3 ...   ...
```

Example of extended CIGAR and the pileup output. (a) Alignments of one pair of reads and three single-end reads. (b) The corresponding SAM file. The '@SQ' line in the header section gives the order of reference sequences. Notably, r001 is the name of a read pair. According to FLAG 163 (=1 + 2 + 32 + 128), the read mapped to position 7 is the second read in the pair (128) and regarded as properly paired (1 + 2); its mate is mapped to 37 on the reverse strand (32). Read r002 has three soft-clipped (unaligned) bases. The coordinate shown in SAM is the position of the first aligned base. The CIGAR string for this alignment contains a P(padding) operation which correctly aligns the inserted sequences. Padding operations can be absent when an aligner does not support multiple sequence alignment. The last six bases of read r003 map to position 9, and the first five to position 29 on the reverse strand. The hard clipping operation H indicates that the clipped sequence is not present in the sequence field. The NM tag gives the number of mismatches. Read r004 is aligned across an intron, indicated by the N operation. (c) Simplified pileup output by SAMtools. Each line consists of reference name, sorted coordinate, reference base, the number of reads covering the position and read bases. In the fifth field, a dot or a comma denotes a base identical to the reference; a dot or a capital letter denotes a base from a read mapped on the forward strand, while a comma or a lowercase letter on the reverse strand.

In SAM, each alignment line has 11 mandatory fields and a variable number of optional fields. The mandatory fields are briefly described in Table. They must be present but their value can be a '*' or a zero (depending on the field) if the corresponding information is unavailable. The optional fields are presented as key-value pairs in the format of TAG: TYPE: VALUE. They store extra information from the platform or aligner. For example, the 'RG' tag keeps the 'read group' information for each read. In combination with the '@RG' header lines, this tag allows each read to be labeled with metadata about its origin, sequencing center and library. The SAM format specification gives a detailed description of each field and the predefined TAGs.

Table: Mandatory fields in the SAM format

No.	Name	Description
1	QNAME	Query NAME of the read or the read pair
2	FLAG	Bitwise FLAG (pairing, strand, mate strand, etc.)
3	RNAME	Reference sequence NAME
4	POS	1-Based leftmost POSition of clipped alignment
5	MAPQ	MAPping Quality (Phred-scaled)
6	CIGAR	Extended CIGAR string (operations: MIDNSHP)
7	MRNM	Mate Reference NaMe ('=' if same as RNAME)
8	MPOS	1-Based leftmost Mate POSition
9	ISIZE	Inferred Insert SIZE
10	SEQ	Query SEQuence on the same strand as the reference
11	QUAL	Query QUALity (ASCII-33=Phred base quality)

Extended CIGAR

The standard CIGAR description of pairwise alignment defines three operations: 'M' for match/mismatch, 'I' for insertion compared with the reference and 'D' for deletion. The extended CIGAR proposed in SAM added four more operations: 'N' for skipped bases on the reference, 'S' for soft clipping, 'H' for hard clipping and 'P' for padding. These support splicing, clipping, multipart and padded alignments. Figure above shows examples of CIGAR strings for different types of alignments.

Binary Alignment/Map format

To improve the performance, we designed a companion format Binary Alignment/Map (BAM), which is the binary representation of SAM and keeps exactly the same information as SAM. BAM is compressed by the BGZF library, a generic library developed by us to achieve fast random access in a zlib-compatible compressed file. An example alignment of 112 Gbp of Illumina GA data requires 116 GB of disk space (1.0 byte per input base), including sequences, base qualities and all the meta information generated by MAQ. Most of this space is used to store the base qualities.

Sorting and Indexing

A SAM/BAM file can be unsorted, but sorting by coordinate is used to streamline data processing and to avoid loading extra alignments into memory. A position-sorted BAM file can be indexed. We combine the UCSC binning scheme and simple linear indexing to achieve fast random retrieval of alignments overlapping a specified chromosomal region. In most cases, only one seek call is needed to retrieve alignments in a region.

SAMtools Software Package

SAMtools is a library and software package for parsing and manipulating alignments in the SAM/BAM format. It is able to convert from other alignment formats, sort and merge alignments, remove PCR duplicates, generate per-position information in the pileup format, call SNPs and short indel variants, and show alignments in a text-based viewer. For the example alignment of 112 Gbp Illumina GA data, SAMtools took about 10 h to convert from the MAQ format and 40 min to index with <30 MB memory. Conversion is slower mainly because compression with zlib is slower than decompression. External sorting writes temporary BAM files and would typically be twice as slow as conversion.

Sequence Profiling Tool

A sequence profiling tool in bioinformatics is a type of software that presents information related to a genetic sequence, gene name, or keyword input. Such tools generally take a query such as a DNA, RNA, or protein sequence or 'keyword' and search one or more databases for information related to that sequence. Summaries and aggregate results are provided in standardized format describing the information that would otherwise have required visits to many smaller sites or direct literature searches to compile. Many sequence profiling tools are software portals or gateways that simplify the process of finding information about a query in the large and growing number of bioinformatics databases. The access to these kinds of tools is either web based or locally downloadable executables.

Typical scenarios of a profiling approach become relevant, particularly, in the cases of the first two groups, where researchers commonly wish to combine information derived from several sources about a single query or target sequence. For example, users might use the sequence alignment and search tool BLAST to identify homologs of their gene of interest in other species, and then use these results to locate a solved protein structure for one of the homologs. Similarly, they might also want to know the likely secondary structure of the mRNA encoding the gene of interest, or whether a company sells a DNA construct containing the gene. Sequence profiling tools serve to automate and integrate the process of seeking such disparate information by rendering the process of searching several different external databases transparent to the user.

Sequerome

Sequerome is a web-based Sequence profiling tool developed by the Bioinformatics and Computational Biosciences Unit (BCBU]) at the Georgetown University. This tool provides a unique and useful functionality of profiling the entire BLAST report by linking it to a panel of servers that perform advanced sequence manipulations in a tabbed browsing environment.

Biologists often perform sequence homology studies using the Basic Local Alignment Search Tool BLAST. The output of the results interface provides a very limited access to sequence analysis tools, such as restriction enzyme maps for DNA sequences and secondary structure prediction for protein sequences. This is further compounded by inability to navigate smoothly between different sequence alignment records. Such needs are better met by a single interface that would allow a user to directly link each of the sequences from alignment reports to different domains/servers offering analysis and manipulation options. Sequerome, a three tiered Java based tool, was developed to meet this end by acting as a front-end to BLAST queries and provide simplified access to web-distributed resources for protein and nucleic acid analysis. Despite the limited scope offered at present, the application is likely to mushroom the growth of related software tools that expand in their scope and coverage of molecular biology tools.

Since its inception in 2005 the tool has been featured in the Science (journal) and officially linked to many Bioinformatics portals around the globe.

Some of the salient features of Sequerome include

- Profiling Sequence alignment reports from BLAST by linking the results page to a panel of third party services.

- Tabbed browsing allowing user to come back earlier operations, visit third party services to perform customized sequence manipulations.

- One-box any-format sequence input. Alternate options for sequence input including visiting third party sites.

- Cached storage of input sequences and retrieval.

- Three pane browsing environment allowing simultaneous input and analysis of multiple sequences.

- Archival options on top of each icon, for results from each pane.

The software application can be accessed directly as a web-server at its Homepage. The homepage shows three panels, viz. Query pane, Results pane and the Search History pane. The user may resize these panes to perform parallel actions in any of these panes. So, in a single browser one could be running parallel BLAST searches on different sequences, analyzing them or viewing the restriction digests for each document of a BLAST result.

Query pane (top panel): Each browser session intends to perform without asking too many questions at the outset. The user has to just dump in the sequence in the Query pane, and BLAST the sequence right away under standard parameters. Experienced users have a choice to perform further special operations under the advanced options. Some of features include selection of specific databases to BLAST from, *upload* facility to work with FASTA files stored in individual computers, sequence retrieval using NCBI IDs and visit any user-defined URL to *drag-N-drop* the sequences. Alternatively the user can also perform a variety of other actions including Sequence manipulation, analysis, and alignment using popular existing tools available in the web. The One-box any-sequence, takes input in any format (FASTA, with or without spaces/numbers). Alerts also exist to warn wrong selection of choices (DNA/RNA/Protein). Results obtained from 'sequence manipulation' e.g. translation, can be further carried on to do further BLAST analysis while preserving the history of the earlier search.

Results Pane (bottom left panel): Sequerome directly queries the input sequence against a variety of databases/tools ('popular public domains' and 'privately hosted services') including BLAST, PDB, REBASE and others, and generates outputs that are intuitive and easily comprehensible. Instant access to various analysis tools, (including viewing a 3D structure-viewer from a PDBid), is provided as separate command buttons to analyze every record from a BLAST report before making a final selection. This is in contrast to the servers that provide plain links to the various resources leaving the user to once feed in the sequence for analysis. In future this could evolve into provision of user-defined analysis tools for every generic analysis e.g. Cn3D, Deepview or Rasmol as structure viewers. In case of results from a Protein BLAST, PDBids are displayed prominently in appropriate cases next to the BLAST record, so that the structure of the molecule with a match can be viewed directly (with an already downloaded version of molecular structure viewer e.g 'Cn3D', 'Rasmol'. Once the BLAST report is displayed on the Results pane, the user can to directly perform a quick analysis on any of the BLAST hits using a series of command buttons that are linked to the respective servers/ sites. Most of the results from third party servers can be viewed directly in the Results pane without opening up as many browsers e.g. ORF prediction, Protparam.

Search history Pane (bottom right panel): One of the key features of a profiling an input sequence data is to store, retrieve and effectively combine and re-use the older inputs. These can be further enhanced if there is retrieval options for each of the operations performed. The bottom right panel in the browser does this while also storing all the input sequences entered earlier. Thus the browser lends an environment to carry out tabbed browsing; an attraction to all those tired of operating multiple browser sessions at a time. For each of the icons linking to the stored results, the user has a choice of archiving them, including print, save and mail options. These can be seen as small colored pictures on top of each icon. This region is less bug-free and one might encounter glitches in the archival options.

Implementation

Sequerome has a three-tiered architecture that uses Java servlet and Server Page technologies with Java Database Connectivity| Java database connectivity (JDBC), making it both server and platform-independent. Sequerome is compatible with essentially all Java-enabled, graphical browsers but is better accessed using Internet Explorer and can be run on most operating systems equipped with a Java Virtual Machine (JVM) and Jakarta Tomcat server. End-users have to download plugins for viewing structure of molecules from PDB e.g. Cn3D, Rasmol, SwissPDB etc.

Bowtie

The Bowtie package enables ultrafast and memory-efficient alignment of large sets of sequencing reads to a reference sequence, such as the human genome. The package contains tools for building indexes of reference genomes and for aligning short reads using the index as a guide. This is the first step of many comparative genomics workflows, including variant detection and digital gene expression. In what follows, the term *read* refers to a short DNA sequence, typically as output by a sequencing instrument. A read may be accompanied by a corresponding string of *quality values*, where each value estimates the probability that the corresponding base was miscalled by the instrument software. The term *subject sequence* refers to the true sequence of the samples from which the reads were drawn. The term *reference sequence* or *reference genome* refers to the sequence to which the subject is to be compared. An example of a commonly used reference sequence is the public human genome assembly GRCh37.

To achieve both speed and memory efficiency, Bowtie aligns reads with the aid of an *index* of the reference genome. The Bowtie index is a refinement of the FM Index, which uses the Burrows-Wheeler Transform to achieve both speed and space efficiency. The user must have built or otherwise obtained an appropriate index before reads can be aligned to the reference. Once an index is built, it can be queried any number of times.

The Basic Protocol describes how to use the bowtie tool to align a collection of reads to a reference genome, assuming an index is available. The Indexing Protocol (Alternate Protocol 1) describes how to build an index for a user-specified reference genome with the bowtie-build tool. The Consensus and SNP Calling Protocol (Alternate Protocol 2) describes how to call consensus sequence, including SNPs, from Bowtie's output using SAMtools. The command line option protocol (Alternate Protocol 3) demonstrates a variety of commonly used alignment options. Finally, the Support Protocols describe how to obtain and install Bowtie software, how to build Bowtie from source code, and how to obtain pre-built indexes from the Bowtie website. Each of these protocols runs in a variety of Unix or Unix-like environments, including Linux, Mac OS X, and Windows.

Basic Protocol

Aligning a Set of Short Reads to a Reference Genome

This protocol uses the bowtie tool to take a collection of short reads and search for each read's best alignment to a reference genome. In typical comparative genomics applications, an alignment is interpreted as a hypothesis for where the read originated in the genome. Once reads are aligned to the reference, the resulting set of alignments becomes a source of evidence for other features of interest, such as DNA sequence differences or gene expression measurements.

Bowtie has many command-line options that allow the user to tune its behavior according to the inputs and the desired result. This protocol describes a few commonly used options and gives some advice regarding how other options should be set.

The first example uses an *E. coli* index that comes with the Bowtie package. Alternate Protocol 1 describes how to build an index from a user-specified reference genome.

All Bowtie options are specified on the command line when invoking Bowtie.

Option	Description
Alignment policy	
−v <int>	Only alignments with up to <int> mismatches in the entire alignment are considered "valid." −v is mutually exclusive with −n. Default: −n mode is enabled and −v mode is disabled.
OR:	
−n <int>	Only alignments with up to <int> mismatches in the seed portion of the alignment are considered "valid." −n is mutually exclusive with −v. Default: 2.
−l <int>	When −n is specified, −l sets the length of the seed portion of the alignment to <int>. Default: 28.

Option	Description
–e <int>	When n is specified, –e sets the limit on the sum of quality scores at the mismatched positions. Default: 70.
Reporting policy	
–k <int>	If Bowtie finds multiple valid alignments for a read, only report the first <int> alignments found. Default: 1.
–m <int>	If Bowtie finds more than <int> valid alignments for a read, suppress all alignments for the read. Default: no limit.
–M <int>	If Bowtie finds more than <int> valid alignments for a read, suppress the alignments but report one of them at random. Default: no limit.
Other options	
–S	Output alignments in SAM format.
–u <int>	Stop and exit after the first <int> reads have been processed.

Descriptions of important command line options for the bowtie alignment tool.

Necessary Resources

Hardware

A Unix, Linux, Mac OS X or Windows 32-bit workstation.

Software

Bowtie 0.12.5 package.

Files

A set of one or more files containing the reads. Reads can be in the FASTA format, FASTQ format (Cock et al., 2009), or in a raw one-sequence-per-line format. The reads used in this example are a set of simulated reads included in the Bowtie package.

A set of index files containing the index of the reference genome. The index file format is unique to Bowtie, and FASTA formats are converted to this format using the bowtie-build tool. The index files used for this example encode the whole genome of *E. coli* strain 536. These files are included in the Bowtie package.

Running Bowtie Interactively from the Command Line

1a   Change to the directory containing Bowtie. For instance, if Bowtie is installed in a directory named

/Users/joe/software/bowtie, type:

- o   cd /Users/joe/software/bowtie

- o   *The example index files are located in the* indexes *subdirectory, and reads are located in the*reads *subdirectory.*

2a   Run Bowtie on the example *E. coli* data with the –u 10 command line option:

./bowtie –u 10 indexes/e_coli reads/e_coli_1000.fq

When invoking Bowtie, the user specifies an index file "basename," i.e. the prefix shared by all the index files (indexes/e_coli in this case), and the path(s) to the FASTQ read files to be aligned (reads/e_coli_1000.fq in this case). All other arguments to Bowtie are interpreted as options.

This command aligns all reads in the specified FASTQ file to the E. coli 536 reference genome, using the specified index as a guide. The –u 10 option causes Bowtie to exit after processing the first 10 reads from the file.

3a   Examine and interpret the output, also shown in Figure .

```
r0      -       gi|110640213|ref|NC_008253.1|  3658049 ATGCTGGAATGGCGATAGTTGGGTGGGTATCGTTC\
        45567778999:9;;<===>?@@@@AAAABCCCDE  0       32:T>G,34:G>A
r1      -       gi|110640213|ref|NC_008253.1|  1902085 CGGATGATTTTTATCCCATGAGACATCCAGTTCGG\
        45567778999:9;;<===>?@@@@AAAABCCCDE  0
r2      -       gi|110640213|ref|NC_008253.1|  3989609 CATAAAGCAACAGTGTTATACTATAACAATTTTGA\
        45567778999:9;;<===>?@@@@AAAABCCCDE  0
r5      +       gi|110640213|ref|NC_008253.1|  4249841 CAGCATAAGTGGATATTCAAAGTTTTGCTGTTTTA\
        EDCCCBAAAA@@@@?>===<;;9:99987776554  0
r7      +       gi|110640213|ref|NC_008253.1|  4086913 GCATATTGCCAATTTTCGCTTCGGGGATCAGGCTA\
        EDCCCBAAAA@@@@?>===<;;9:99987776554  0
r8      +       gi|110640213|ref|NC_008253.1|  2679194 GGTTCAGTTCAGTATACGCCTTATCCGGCCTACGG\
        EDCCCBAAAA@@@@?>===<;;9:99987776554  0       14:A>T,33:C>G
r9      -       gi|110640213|ref|NC_008253.1|  2430559 GCCTGTTCGGCGTTGAGGGTAATGAAATCATCGCC\
        45567778999:9;;<===>?@@@@AAAABCCCDE  0
# reads processed: 10
# reads with at least one reported alignment: 7 (70.00%)
# reads that failed to align: 3 (30.00%)
Reported 7 alignments to 1 output stream(s)
```

Sample Bowtie session. The last few lines are written to the "standard error" file handle and convey summary information about the run. The other lines are written to the "standard out" file handle and show valid, reportable alignments found by Bowtie. In this and subsequent figures, the backslash character is used to indicate that some long lines are wrapped.

Note that in practice, users rarely need to examine the alignment output manually. Rather, users will typically redirect Bowtie's output to a file or directly to another program for downstream processing.

The output consists of two sets of lines, those printed to the "standard out" file handle and those printed to the "standard error" file handle. Each line printed to "standard out" is a valid, reportable alignment found by Bowtie. A valid alignment is one that satisfies the alignment policy specified in the –n, –l, –e or –v options. A reportable alignment is one that is not suppressed by any other option, such as the –k, –m, or –M options. Lines printed to "standard error" contain summary information about the entire alignment run, e.g., the total number of reads that were processed and the percentage of total reads for which Bowtie found at least one alignment.

The format for an alignment consists of 8 fields, separated by tabs. The fields, from left to right, are:

    a. Read name.

    b. Reference strand aligned to ("+" denotes forward strand, "–" denotes reverse strand).

    c. Name of reference sequence aligned to.

    d. Offset of the leftmost position on the forward reference strand covered by the alignment.

    e. Read sequence aligned (or its reverse complement, if the read aligned to the reverse strand).

f.   Quality sequence aligned (or its reverse, if the read aligned to the reverse strand).

g.   A string representing the differences between the read and the corresponding characters of the reference genome. Each difference is described in a comma-separated field.

4a   To obtain alignment results in SAM format, a file format recognized by SAMTools and many other sequence analysis packages, add the –S option to the bowtie command-line:

```
./bowtie -S -u 10 indexes/e_coli reads/e_coli_1000.fq
```

This will produce output similar to the earlier example, except that the alignments printed to standard output will be in SAM format.

```
@HD     VN:1.0  SO:unsorted
@SQ     SN:gi|110640213|ref|NC_008253.1|  LN:4938920
@PG     ID:Bowtie       VN:0.12.5       CL:"./bowtie -S -u 10 indexes/e_coli reads/e_coli_1000.fq"
r0      16      gi|110640213|ref|NC_008253.1|  3658050  255     35M     *       0       0\
        ATGCTGGAATGGCGATAGTTGGGTGGGTATCGTTC     45567778999:9;;<===>?@@@@AAAABCCCDE     XA:i:0\
        MD:Z:0G1T32     NM:i:2
r1      16      gi|110640213|ref|NC_008253.1|  1902086  255     35M     *       0       0\
        CGGATGATTTTTATCCCATGAGACATCCAGTTCGG     45567778999:9;;<===>?@@@@AAAABCCCDE     XA:i:0\
        MD:Z:35 NM:i:0
r2      16      gi|110640213|ref|NC_008253.1|  3989610  255     35M     *       0       0\
        CATAAAGCAACAGTGTTATACTATAACAATTTTGA     45567778999:9;;<===>?@@@@AAAABCCCDE     XA:i:0\
        MD:Z:35 NM:i:0
r3      4       *       0       0       *       *       0       0       AAAATTTGTGCCTGGATGGCCTGAGTACCNANTAC\
        EDCCCBAAAA@@@@?>===<;;9:99987776554     XM:i:0
r4      4       *       0       0       *       *       0       0       GCAGAGCAGTTGCTAGAAANNNNNTTGAAGAGGTT\
        EDCCCBAAAA@@@@?>===<;;9:99987776554     XM:i:0
r5      0       gi|110640213|ref|NC_008253.1|  4249842  255     35M     *       0       0\
        CAGCATAAGTGGATATTCAAAGTTTTGCTGTTTTA     EDCCCBAAAA@@@@?>===<;;9:99987776554     XA:i:0   MD:Z:35\
        NM:i:0
r6      4       *       0       0       *       *       0       0       GGCAGTGATGCAACTGCCCGTTATCAACAGNCNCT\
        EDCCCBAAAA@@@@?>===<;;9:99987776554     XM:i:0
r7      0       gi|110640213|ref|NC_008253.1|  4086914  255     35M     *       0       0\
        GCATATTGCCAATTTTCGCTTCGGGGATCAGGCTA     EDCCCBAAAA@@@@?>===<;;9:99987776554     XA:i:0   MD:Z:35\
        NM:i:0
r8      0       gi|110640213|ref|NC_008253.1|  2679195  255     35M     *       0       0\
        GGTTCAGTTCAGTATACGCCTTATCCGGCCTACGG     EDCCCBAAAA@@@@?>===<;;9:99987776554     XA:i:1\
        MD:Z:14A18C1    NM:i:2
r9      16      gi|110640213|ref|NC_008253.1|  2430560  255     35M     *       0       0\
        GCCTGTTCGGCGTTGAGGGTAATGAAATCATCGCC     45567778999:9;;<===>?@@@@AAAABCCCDE     XA:i:0   MD:Z:35\
        NM:i:0
# reads processed: 10
# reads with at least one reported alignment: 7 (70.00%)
# reads that failed to align: 3 (30.00%)
Reported 7 alignments to 1 output stream(s)
```

Sample Bowtie session when SAM output mode is enabled. The last few lines are written to the "standard error" file handle and convey summary information about the run and are not part of the SAM format. The first few lines beginning with "@" are the SAM header. The remaining lines are alignments.

5a   When aligning a large number of reads, the user will wish to capture the standard output to a file rather than to have it appear on the screen. The user can do this by redirecting standard output as shown here:

```
./bowtie -S indexes/e_coli reads/e_coli_1000.fq > alignments.sam
```

This will align all reads contained in the E. coli FASTQ file and save them in SAM format to the file named alignments.sam.

Alternate Protocol

Building an Index For a Set of Reference Sequences

This protocol uses the bowtie-build tool to take a collection of FASTA files for a reference genome and generate a collection of index files. Index files can then be used by bowtie to align reads to the reference genome. The same set of index files can be used across multiple runs of bowtie.

Bowtie-build has command-line options that allow the user to tune its behavior, but this protocol will use the defaults.

Necessary Resources

Hardware

A Unix, Linux, Mac OS X or Windows 32-bit workstation.

Software

Bowtie 0.12.5 package.

1. Download the two compressed FASTA files corresponding to the human sex chromosomes (chrX.fa.gz and chrY.fa.gz) from the UCSC FTP site at .

2. In the same directory, use gunzip to decompress the compressed FASTA files that were just downloaded:

    gunzip *.gz

3. Run bowtie-build to build an index of the two human sex chromosomes:

    bowtie-build chrX.fa,chrY.fa human_xy

    This builds an index consisting of the sequences in the chrX.fa *and* chrY.fa *files and stores the results in a set of six files with prefix "human_xy". If the* bowtie-build *executable is not in the search path, specify the full path to* bowtie-build *instead. This command typically takes about ten or fifteen minutes.*

    In this example each FASTA file contains one sequence. FASTA files with multiple sequences ("multi-FASTA" files) are also permitted, in which case all sequences in the multi-FASTA file are included as separate sequences in the index.

4. Confirm that the output files were created:

    ls –l human_xy*

    The output should be similar to figure below. Note the index consists of 6 distinct files with the same prefix (human_xy), *and suffix* (.ebwt).

    ```
    -rw-r--r--  1 user  group  54696183 May 28 15:24 human_xy.1.ebwt
    -rw-r--r--  1 user  group  22094272 May 28 15:24 human_xy.2.ebwt
    -rw-r--r--  1 user  group       377 May 28 15:20 human_xy.3.ebwt
    -rw-r--r--  1 user  group  44188532 May 28 15:20 human_xy.4.ebwt
    -rw-r--r--  1 user  group  54696183 May 28 15:27 human_xy.rev.1.ebwt
    -rw-r--r--  1 user  group  22094272 May 28 15:27 human_xy.rev.2.ebwt
    ```

Listing of files output by bowtie-build after indexing the human sex chromosomes.

5. Inspect the index using bowtie-inspect:

    bowtie-inspect –s human_xy

    The output should be similar to Figure *. Among other things, the output tells the user that the index is not for colorspace alignment (first line), and shows the names and lengths of the two reference sequences included in the index (last two lines).*

```
Colorspace       0
SA-Sample        1 in 32
FTab-Chars       10
Sequence-1       chrX 155270560
Sequence-2       chrY 59373566
```

Output of bowtie-inspect when inspecting the index consisting of the two human sex chromosomes.

Understanding Results

The purpose of alignment is to determine reads' *point of origin* with respect to the reference genome. Once points of origin are identified, downstream tools use that information, for example, to characterize differences between the subject and reference genome (e.g. when calling SNPs), or to relate the reads to annotations defined with respect to the reference genome (e.g. for digital gene expression). Alignment programs, together with appropriate reference sequences, serve this purpose because genomes of individuals of the same species tend to be highly similar. For example, two humans typically have on the order of 3–4 million single-nucleotide differences between them out of a total of 3 billion bases. However comparative strategies also have inherent drawbacks that should be kept in mind when interpreting Bowtie results.

Repetitive Genome Content Affects Results

Some genomes, including the human genome, have substantial repetitive content, i.e. sub-sequences that appear multiple times throughout the genome. Repeats come in several forms (e.g. simple repeats, tandem repeats, segmental duplications, interspersed repeats), and arise via various biological processes (e.g. slipped strand mispairing or retrotransposition). Repeats also affect alignments because reads originating from repetitive portions of the genome are difficult or impossible to *unambiguously* assign to a point of origin. Reads from repeats will tend to have many "valid" alignments, with no strong basis for preferring one over the others. Paired-end reads mitigate but do not necessarily eliminate this problem. Repetitive alignments in turn affect downstream analyses. For instance, if ambiguous alignments are included in the output from Bowtie, a SNP could yield false positives and false negatives purely owing to the repeat structure.

The simplest way to deal with alignment ambiguity is to use Bowtie's –m and –M options to filter out and/or annotate ambiguous evidence as such. When ambiguous alignments are annotated in this way, downstream tools can choose to discount or ignore evidence from ambiguous alignments as appropriate. With the –moption, the user specifies a limit whereby alignments for reads that align to a number of locations exceeding the limit are excluded from the output. For instance, to suppress all alignments for reads that align to more than 3 locations on the reference, the user specifies –m 3. The –M option is similar to the –moption except that one random alignment from among those found is reported, but is marked as being ambiguous.

Sequence Differences between Subject and Reference Affect Results

Though human genomes are close to 99.9% similar at the sequence level, the differences can have a significant effect on alignment. For instance, if the subject genome contains a cluster of 3 single-nucleotide variants within 10 bases of each other, and Bowtie is run with the −v 2 option (i.e. only alignments with up to 2 mismatches are valid), Bowtie is very unlikely to report the correct alignment for a read whose true point of origin spans the three variants. A similar example involves gaps, if the subject genome has a gap with respect to the reference genome (or vice versa), Bowtie is unlikely to report the correct alignment for a read whose true point of origin spans the gap. These potential effects must be taken into account when interpreting output from Bowtie.

Sub-disciplines of Bioinformatics

Bioinformatics is a vast subject that branches out into a number of significant sub-disciplines such as integrative bioinformatics, neuroinformatics, glycoinformatics, bioimage informatics, neighbor joining, etc. These sub-disciplines have been thoroughly discussed in this chapter.

Integrative Bioinformatics

Integrative Bioinformatics deals with the development of methods and tools to solve biological problems as well as providing a better understanding or new knowledge about biochemical phenomena by means of data integration and computational experiments. Current high-throughput technologies such as NMR, mass spectrometry, protein/DNA chips, gel electrophoresis data, Yeast Two-Hybrid, QTL mapping, and NGS generate large quantities of high-throughput data. The challenge of Integrative Bioinformatics is to capture, model, simulate, integrate, and analyze this huge amount of data in addition to the data represented by hundreds of biological databases and thousands of scientific journals. The data needs to be integrated and made available in a consistent way to provide new and deeper insights into complex biological systems. Molecular biology produces this volume of data based on high-throughput technologies. One characteristic of this data is exponential growing. Therefore, storing and analysis of this molecular and cellular data essentially uses methods and concepts of Bioinformatics. Currently, there are more than 2,000 database and information systems available via the Internet, which represent this molecular data. Every year new molecular databases and information systems which can be used via the Internet crop up.

Integrative Bioinformatics Approaches for Identification of Drug Targets in Hypertension

High blood pressure or hypertension is an established risk factor for a myriad of cardiovascular diseases. Genome-wide association studies have successfully found over nine hundred loci that contribute to blood pressure. However, the mechanisms through which these loci contribute to disease are still relatively undetermined as less than 10% of hypertension-associated variants are located in coding regions. Phenotypic cell-type specificity analyses and expression quantitative trait loci show predominant vascular and cardiac tissue involvement for blood pressure-associated variants. Maps of chromosomal conformation and expression quantitative trait loci (eQTL) in critical tissues identified 2,424 genes interacting with blood pressure-associated loci, of which 517 are druggable. Integrating genome, regulome and transcriptome information in relevant cell-types could help to functionally annotate blood pressure associated loci and identify drug targets.

Elevated blood pressure (BP) or hypertension is a heritable chronic disorder, considered the single largest contributing risk factor in disease burden and premature mortality. High systolic and/

or diastolic BP reflects a higher risk of cardiovascular diseases. Genome-wide association studies (GWAS) have found association of 905 loci to BP traits (systolic - SBP, diastolic - DBP and pulse pressure -PP) to date. The use of larger sample sizes has helped to identify additional variants, as demonstrated by the most recent study including over 1 million people that has identified 535 novel BP loci. Still, this collective effort thus far has not entirely elucidated the complete genetic contribution to BP, estimated to be approximately 50–60%.

To add to this complexity, 90.7% of the 905 BP-associated index variants are located in intronic or intergenic regions). Causal variants are also difficult to pinpoint because of linkage disequilibrium (LD). There is now vast evidence that non-coding variants associated with disease interrupt the action of regulatory elements crucial in relevant tissues for that particular disease. BP loci are not only linked to cardiovascular disease but also to other diseases, suggesting that BP-associated variants can result in a wide range of phenotypes. Tissue specificity of genetic loci may be relevant for organ specific disease progression. For example, variants altering expression in heart may more likely affect disease progression through heart-mediated processes rather than kidney-mediated processes, and some patients may suffer of left ventricular hypertrophy while others may develop nephropathy. Thus, investigating the influence of BP variants in critical cell-types is essential in understanding disease risk and biology, and assessing the possible translation of an associated locus into a drug target. The public availability of regulatory annotations in several tissues by projects such as ENCODE, Roadmap and GTEx has enabled integration of epigenetic modifications, expression quantitative trait loci (eQTLs) and –omics information with GWAS data. Integrative approaches are useful for prioritizing genes from known GWAS loci for functional follow-up, detecting novel gene-trait associations, inferring the directions of associations, and potential druggability.

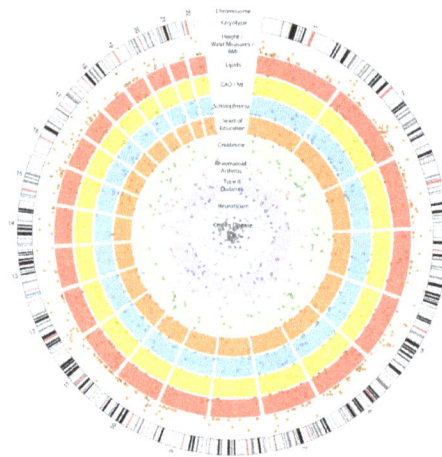

Circos plot showing the 10 traits from the GWAS catalog with the largest number of loci also associated to BP, as identified by PhenoScanner at $p < 0.05$ (Supplemental Methods). The outer ring represents the genomic/chromosomal location (hg19). The following inner rings show the associations to different traits. Beige: body measurements (height, body mass index (BMI), weight, waist/hip ratio, hip circumference, waist circumference. $N = 358$). Red: lipids (high-density lipoprotein (HDL), low-density lipoprotein (LDL), triglycerides, total cholesterol. $N = 226$). Yellow: coronary artery disease (CAD)/myocardial infarction (MI) ($N = 206$). Blue: schizophrenia ($N = 135$). Orange: years of education attendance ($N = 101$). Light green: creatinine ($N =$

88). Light pink: rheumatoid arthritis ($N = 78$). Purple: type II diabetes ($N = 73$). Light turquoise: neuroticism ($N = 69$). Light grey: Crohn's disease ($N = 67$).

Here we summarize the advances made in recent years towards unraveling the mechanisms of non-coding BP variants in disease progression with the resources mentioned above. We focus on integrative approaches that aim to prioritize BP-associated SNPs located in regulatory regions of the genome for follow-up studies.

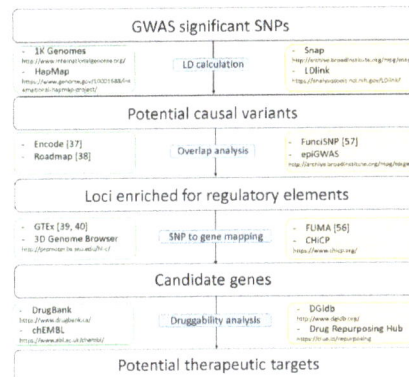

Figure above shows analytical steps that can be followed for variant prioritization and translation of association to a potential drug target. Each step is accompanied by examples of publicly available data (green boxes on the left) and tools (yellow boxes on the right) that can be used.

Integrative Approaches using Omics Data

Remarkable advances have been made recently towards a better comprehension of BP genetics, the biology of disease and translation towards new therapeutics, boosted by the widespread application of high-throughput genotyping technologies. At the same time, most BP-associated variants are non-coding, making the conversion of statistical associations into target genes a great challenge. SIFT, PROVEAN, PolyPhen, CONDEL and more recently CADD are scoring algorithms developed for predicting the effect of amino acid changes. Only 98 out of the 905 lead BP-associated SNPs reflect a CADD score above 12.37, a threshold suggested by Kicher et al. as deleterious. However, the causal variant inside the locus might reflect a different CADD score than the lead SNP, and pinpointing the mechanisms disturbed by the variation remains a challenge.

New strategies that make use of regulatory annotations in disease-relevant tissues have greatly expanded our ability to investigate the processes involved in BP. In particular, annotation of histone modifications and regions of open chromatin allow the identification of active transcription in specific-cell types. Similarly, maps of DNA variants affecting expression in a cell-type specific manner will be integral in BP loci interpretation. A list of cardiovascular-related cell-types researched by the ENCODE Project is presented by Munroe et al. Such data can be integrated with GWAS results using bioinformatics tools. For instance, FUMA provides extensive functional annotation for all SNPs in associated loci and annotates the identified genes in biological context. FunciSNP investigates functional SNPs in regulatory regions of interest. Ensemble's Variant Effect Predictor (VEP) determines the effect of variants on genes, transcripts, and protein sequence, as well as regulatory regions, also outputting SIFT, Polyphen and CADD scores for each variant, among other information. Although such integrative tools are useful for variant prioritization and interpretation, not all take into consideration tissue specificity aspects. RegulomeDB, for example, is a database that annotates SNPs with known and predicted regulatory elements in the intergenic regions of the human genome, calculating a score that reflects its evidence for regulatory potential. However, the scoring procedure can only be performed across all available tissue types. In addition, several databases containing a broad range of tissues were made publicly available since the last update of RegulomeDB, that could be included in the tool. Together, these resources have been useful in prioritizing genes and variants in associated loci for functional follow-up experiments in many post-GWAS analyses, and can be implemented in interpretation of BP-associated loci.

Transcription Regulation: Histone Modifications and Open Chromatin

As genomic coordinates of active regulatory elements may be mapped using unique functions of chromatin, the characterization of chromatin changes in the genome in specific cell-types can be used to identify DNA variants disturbing active regulatory elements. The four core chromatin histones, H2A, H2B, H3 and H4, can suffer posttranslational modifications, such as acetylation or methylation. These histone modifications indicate active (euchromatin) or repressed (heterochromatin) chromatin structure, defining regulation and gene transcription. Acetylation of histones H3 and H4, and H3 methylation at Lys4 (H3K4me3), for instance, correlate with gene transcription, whereas methylation at Lys9 correlates with gene silencing. These modifications provide a robust readout of active regulatory positions in the genome, and have been employed for annotation in several studies. Histone modifications influencing arterial pressure have been observed in many tissues, including vascular smooth muscle . An updated phenotypic cell-type specificity analysis of the 905 BP loci using H3K4me3 mark in 125 tissues is shown in figure below. The most significant cell-types are cardiovascular-related (Supplemental Methods, Table). Other tissues with high rank in specificity are smooth muscle, fetal adrenal gland, embryonic kidney cells, CD34 and stem-cell derived CD56 +mesoderm cultured cells. These results are consistent with analyses using DNase I hypersensitivity sites (DHSs), which indicate likely binding sites of transcription factors. These results add more evidence that BP loci are enriched on regions of open chromatin, regulating transcription in a broad range of tissues.

Ranked tissues after phenotypic cell-type specificity analysis of 905 BP SNPs using 125 H3K4me3 datasets on human tissue (Supplemental Methods).

Methylation

In addition to histone modifications that promote transcription, BP loci have also been studied for their enrichment on DNA methylation, known to have the opposite regulatory effect. The methylation of CpG sites, presented by CpG islands in promoters, affects binding of transcription factors, resulting in gene silencing. Abnormal CpG methylation is found in hypertension, and in many other complex diseases. Recently, Kato et al. identified a ~2 fold enrichment associating BP variants and local DNA methylation. The study also demonstrates that DNA methylation in blood correlates with methylation in several other tissues. These observations add to previous indications on the function of DNA methylation in regulating BP.

Measuring the Impact of BP Risk Alleles on Gene Expression: eQTLs

Expression quantitative trait loci (eQTL) are regions harbouring nucleotides correlating with alterations in gene expression. Linking transcription levels to complex traits has been a follow-up step adopted by many studies, driven by the increase in available data of expression patterns across tissues and populations. Warren et al. found that 55.1% of their identified BP-associated loci have SNPs with eQTLs in at least one tissue from GTex repository, with arterial tissue most frequently observed (29.9% of loci had eQTL in aorta and/or tibial artery). A great enrichment of eQTLs in artery was also observed by Evangelou et al., who identified 92 novel loci with eQTL enrichment in arterial tissue and 48 in adrenal tissue. In summary, these studies also suggest that BP loci exert a regulatory effect mostly in vascular and cardiac tissues.

Finding the Targets: Chromosome Confirmation Capture Techniques

Mapping variation to target genes is one of the greatest challenges in the post-GWAS era, and different strategies have been developed to this end. One approach is the use of chromosome confirmation capture [3C], 4C, Hi-C]. These techniques capture chromosome interactions, resulting in networks of interacting genetic loci.

Warren made use of this resource to investigate the target genes of non-coding SNPs, using Hi-C data from endothelial cells (HUVECs). Distal potential genes were found on 21 loci, and these genes were enriched for regulators of cardiac hypertrophy in pathway analysis Kraja et al. also explored long-range chromatin interactions using endothelial precursor cell Hi-C data, finding the link between an associated loci and a gene known to affect cell growth and death. More recently, Evangelou used chromatin interaction Hi-C data from HUVECs, neural progenitor cells (NPC), mesenchymal stem cells (MSC) and tissue from the aorta and adrenal gland to identify distal affected genes. They found 498 novel loci that contained a potential regulatory SNP, and in 484 loci long-range interactions were found in at least one cell-type.

A list of human HiC data available on BP relevant tissues is presented. An updated version of variant to gene mapping making use of this chromatin conformation data is shown in Table . Promoter regions of 1,941 genes were found to interact with the 27,649 candidate SNPs (905 BP associated SNPs and vicinity). Integration with eQTL data on relevant tissues confirmed 209 of the genes mapped, and added additional 483 genes. One main goal of understanding biological mechanisms of GWAS associations and affected genes is to be able to therapeutically target them. Assessment of the druggability of a BP-associated locus depends on several factors, but overlap of these results with a recent effort on druggability suggests that 517 of these 2,424 genes are druggable, and 35 mapped genes are also predicted to interact with common drugs for treatment of hypertension, Supplementary Methods). Interestingly, 1,774 of the genes mapped are physically located outside BP-associated loci. These results support the hypothesis that BP GWAS loci act on tissue specific regulatory gene networks. Importantly, they also show that the use of long range chromatin interaction maps can reliably identify target genes even outside the risk locus.

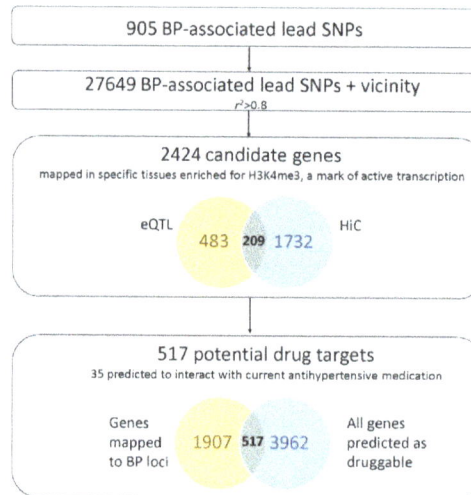

Results of our integrative approach.

Neuroinformatics

Neuroinformatics as a field that includes building databases and tools for understanding the nervous system was initiated in the early 1990s. During the subsequent decade development started on dealing with the complex types of data that characterize studies of the nervous system. In 2000, a paper suggested that web portals (databases) were needed to catalog the burgeoning number of databases and tools. Around 2004, a database (SfN (Society for Neuroscience). emerged to satisfy this need, and a more ambitious implementation of the approach is currently under development.

It may facilitate understanding the complexity of neuroinformatics by first discussing a related field, bioinformatics. The first protein data collections were made in the late 1970's; reviewed in . Investigators realized that the computer was an essential tool to keep track of either the series of letters (e.g. GCAT) that represented the base sequence that makes up the nucleic acids or the sequence of amino acids that make up proteins. Databases for genes, European Molecular Biology laboratory's EMBL-Bank and the National Center for Biotechnology Information's GenBank, were launched in the early 1980s.

Neuroinformatics Compared to Traditional Bioinformatics

Traditional bioinformatics is the field that encompasses comparing and databasing the genome (DNA), and related molecules (RNA, proteins), and also modeling the structure and function of existing and new (designed) proteins. We use the term traditional bioinformatics to acknowledge that the explosion of activity in bioinformatics has grown beyond the origin of bioinformatics. Bioinformatics can be defined most generally (although not all investigators choose to do so) as all combinations of biology (life sciences) and informatics (computer and statistical methods). Bioinformatics most general definition would then include neuroinformatics as well as systems biology (that seeks to model all aspects of life) as part of bioinformatics. Several online descriptions further elaborate or nuance this broad view of bioinformatics. Bioinformatics' original focus was very specific, profound, and important. We return to elaborating traditional bioinformatics since it is helpful to understand neuroinformatics (and other recent developments in bioinformatics). In early bioinformatics the essential data is the list of letters that make up the sequence. Other example attributes are the species that the sequence is from, chromosomes (if the sequence is DNA) that the sequence belongs to, and the names of the sequence. It was critical to develop tools to allow comparisons between genes, thereby allowing statements to be made about how similar the genes are across alleles, related genes (most genes are thought to be formed from other ancestral genes by duplication and subsequent mutation) and across species. Traditional bioinformatics is comparatively simpler than the newer extensions to bioinformatics such as neuroinformatics because the basic data type in traditional bioinformatics is the sequence. Conceptually traditional bioinformatics consists of sequence oriented databases plus tools to search (on one database, or across databases), to compare (either on-line databases or download-able software), and increasingly to model the molecules related to the sequences.

In contrast, neuroscience data is diverse and heterogeneous. In each subfield of neuroscience, however, there is often an associated primary type of data and neuroinformatics tools to store in databases, to search, to compare, and increasingly to model the physical system. The cross-disciplinary nature of neuroinformatics has required collaboration of teams of scientists with mapping efforts and/or hypothesis-driven goals. Cultural issues are present as a result of these efforts that are new (but not restricted to neuroinformatics).

Neuroinformatics experimental databases may store function and anatomy annotations of nervous system genes, images (acquired with different methods such as structural and functional magnetic resonance imaging (MRI and fMRI), tissue staining at spatial scales from subcellular electron microscope images to tens of centimeters slices though brains of monkeys, optical recordings of voltage and chemical activated dyes, etc.), and atlases of central and peripheral nervous systems. Several projects tackled the difficult problem of open ended data-sharing with heterogeneous data. Experimental and descriptive data, in the open ended case, is documented with meta-data which describes its content and format for successful sharing. Other projects attempted to advance new experimental methods (wavelet analysis in fMRI) and data analysis tools for MRI mapping of brains to surfaces for comparison between brains from the same species (in humans this has medical applications), and also comparisons between different species nervous systems). Further tools allow the recognition of objects within the fMRI data-sets, improve the quality of fMRI images and systems to store and share fMRI data-sets. Other projects were designed to setup databases of in-situ hybridization data to provide a permanent accessible archive for resources in danger of disappearing. Several projects created online atlases and maps of human, macaque (monkey), rat, and mouse brains. Several projects combined databases of images with annotations of gene-expression (within the images).

Theoretical neuroinformatics projects tackled diverse topics: modeling cortical maps, modeling the olfactory bulb and storing computational neuroscience models in a web accessible database, developing new wavelet based and source separation analysis tools for fMRI data, detailed tools for neuromorphological modeling, automatic cortical surface reconstruction mapping and tools for specifying a 2-D coordinate system for the mapping, and creating realistic computational models capable of predicting human auditory responses to a wide range of acoustic stimuli including acoustic trauma.

In conclusion, there is a wide variety of neuroscience databases and tools that support neuroscience and biomedicine. The field of neuroinformatics might be perceived as an extended and elaborated application of analogous tools to those found earlier to be critical for the development of bioinformatics, but applied to a broader, heterogeneous types of data at many levels of function. The depth and breadth of neuroscience information is beginning to be organized through databases of databases. These, and the information in the databases they contain, will aid searching in, comparing, and modeling nervous systems at spatial scales from the level of molecules to behavior, in normal, diseased, or injured humans and animals. Neuroinformatics is becoming essential to neuroscience investigators and clinicians for conducting scientific inquiry, and practicing medicine in our time of rapidly expanding cross-disciplinary knowledge.

Neuroinformatics Examples: Multiple Database Search

Using the National Center for Biotechnology Information (NCBI)

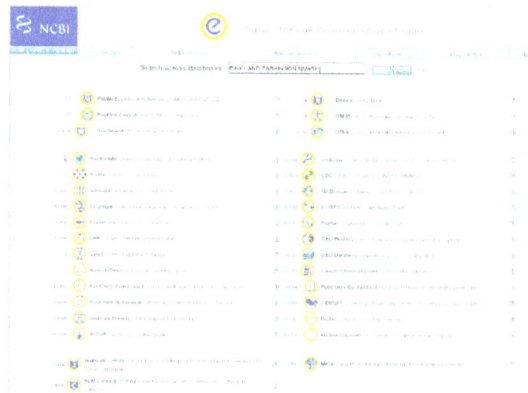

A neuroscience search that demonstrates an overlap in traditional bioinformatics and neuroinformatics starts with an "All database" search at PubMed. This powerful search engine simultaneously searches through 28 databases which includes entries in the diverse topics of literature (articles in journals and books), sequence databases, metadata (descriptions of the data or format in which the data is stored), databases on Journals (themselves, not the articles in them) and vocabulary, and bibliographic data for the National Library of Medicine holdings of books, software, and other resources. This is useful to simultaneously find information about genes and also the literature that describes findings about the gene. A search on the Parkinson's related gene PINK1 results in 114 articles found in PubMed, however using the Medical Subject Heading (MeSH) terms (selected "meaningful words", i.e. words whose definitions are precisely specified, the search "PINK1 AND PARKINSON'S[MeSH]" finds 67 articles which are then known to be relevant to PARKINSON'S disease. The MeSH terms help the user find the items in the databases that have the right context, e.g. the Nucleotide sequence database (includes GenBank) finds 41 hits in the unrestricted

PINK1 search, however, when the PARKINSON'S term is added the number of hits drops to 2 and the reports associated with the entries are targeted to Parkinson's clinical research. This example demonstrated how targeted searches through a collection of databases produces results that are of immediate interest to the neuroscience investigator.

All database search in PubMed. A collection of 23 databases are searched simultaneously for a boolean expression of keywords of interest. Restricting searches with MeSH terms targets the results to relevant topics.

Using the Neuroscience Database Gateway (NDG)

Let us imagine that we are an investigator looking for web sites that contain human brain atlases to reexamine some facet of nervous system anatomy. Starting at the NDG home page click on the "Search" link in the left hand column and enter "Atlas" under categories and "Human" under species and press the "Search" button. If the search terms are not known the "keyword" buttons for each field can be pressed which pop-up a window where multiple search terms can be selected. In the results numbers 15, "Whole Brain Atlas", and 16, "The Navigable Atlas of the Human Brain", are seen by their names to be immediately relevant.

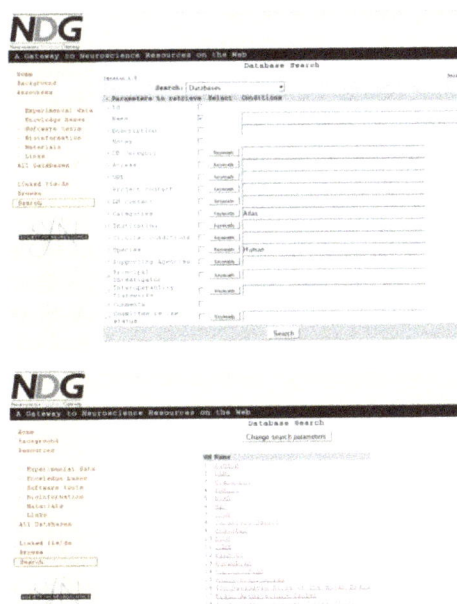

Database search in NDG. The SfN Neuroscience Database Gateway in a sample search for Human Atlases (top). There were 18 results (bottom) of which numbers 15 and 16 (by their names) are suitable for browsing to gain human brain familiarity.

Glycoinformatics

Carbohydrates, often referred to as glycans, differ from other biopolymers such as proteins or nucleic acids in various ways. The number of different monosaccharides that are present in naturally occurring glycans is significantly higher than the number of proteogenic amino acids, or of nucleotides that form DNA or RNA strands. Furthermore, the monosaccharides can be linked to each other in several ways, including the possibility to form branched structures. Another important difference between glycans, on the one hand, and proteins and nucleic acids, on the other hand, is visible in their biosynthesis: DNA, RNA and proteins are synthesized by copying, transcription or translation, respectively, of nucleic acids, whereas carbohydrates are built in a non-template-driven approach by the sequential action of various glycosyltransferases (GT) that add monosaccharides to an existing glycan chain, and by glycoside hydrolases (GH) that remove specific monosaccharides. For this reason there is no technique available to amplify carbohydrates comparable to Polymerase Chain Reaction (PCR) or protein expression systems. Instead, carbohydrates have to be analyzed in physiological amounts. If specific and well-defined glycans are required for experiments such as glycan arrays, they have to be synthesized chemically.

The special features of carbohydrates not only pose problems for their wet-lab analysis but also for computational approaches that deal with carbohydrates. Classical bioinformatics algorithms are developed for linear gene or protein sequences, and thus cannot be applied to branched carbohydrates. Instead, new algorithms that deal with the branching as well as with other special features of carbohydrates, such as microheterogeneity, have to be developed. Furthermore, there are much less primary data on carbohydrates available than, e.g., on proteins, to test or train the algorithms . For these reasons, glycoinformatics as a research area at the intersection of bioinformatics and chemoinformatics has been considered to be lagging behind its sister fields, such as bioinformatics for genomics or proteomics, for a long time. By now, however, glycoinformatics is coming of age and offers a variety of databases and applications that are of use to glycoscientists. Many new resources are still being developed, and efforts for a better integration of existing resources have also been started. Formats and protocols for data exchange have been specified. Recently, the MIRAGE (Minimum Information Related to A Glycomics Experiment) consortium was founded to define checklists for the standardization of experimental glycomics data and meta information. However, there is still no long-term repository of glycan structures available.

Carbohydrate Databases

Glycan Structure Databases

Various databases that collect information on carbohydrates are now available and new resources are still being developed. The individual databases differ in the kind of data that are stored, the number and topicality of entries, the search interfaces, and the way the data are presented to the user. They are of use to glycochemists in several ways. First of all, they provide literature references on specific carbohydrate structures, which are often difficult to find via keyword searches in general literature databases such as PubMed. However, keeping databases up to date with bibliographic references is a time-consuming task that cannot be performed automatically by computer programs because the glycan structures are often encoded graphically within the publication figures. And even when information on glycan chains is given in the text, the notation is often complex, difficult to parse, and may contain ambiguities. Therefore, database users should keep in mind that if a database does not list any reference that, e.g., deals with the synthesis of a specific glycan structure, it does not mean that there is no such reference available: it just might not have been included into the database yet. Aside from providing literature references, carbohydrate databases can also serve glycochemists as a source of information on structures that are potential targets for synthesis. For this purpose resources that feature data such as the biological source, or diseases related to a glycan structure, can be of special interest.

Bioimage Informatics

In the last several decades, numerous biomedical imaging techniques were developed, ranging from the whole organism level (millimeter resolution) down to the single molecule level (nanometer resolution). Some of the most widely used biological imaging methods include confocal or two-photon laser scanning microscopy (LSM), scanning or transmission electron microscopy (EM), etc. Novel imaging techniques such as PALM, STORM, STED that far surpass the resolution of conventional optical microscopes currently can pinpoint the location of individual proteins that are only several nanometers apart. Along with the dramatic advances of many related techniques such as image signal digitization and storage, biological tissue labeling [e.g. green fluorescent proteins (GFP) and enhanced GFP (EGFP), Dronpa, Brainbow combinatorial labeling], the number of biological images (e.g. cellular and molecular images, as well as medical images) acquired in digital forms is growing rapidly. Large bioimage databases such as Allen Brain Atlas and the Cell Centered Database CCDB; are becoming available. These image data could involve (1) two-dimensional (2D) or 3D spatial information, (2) multiple colors which may correspond to various molecular reporters, (3) 4D spatio-temporal information for developing tissues or moving cells, (4) various co-localized biological signals such as mRNA expression levels of different genes or (5) other screening experiments related to RNA interference (RNAi), chemical compounds, etc. Analyzing these images is critical for biologists to seek answers to many biological problems, such as differentiating cancer cell phenotypes, categorization of neurons, etc.

The deluge of complicated biological and biomedical images poses significant challenges for the image computing community. As a natural extension of the existing biomedical image analysis field, an emerging new engineering area is to develop and use various image data analysis and informatics techniques to extract, compare, search and manage the biological knowledge of the respective images. This new field can be called bioimage informatics. However, due to the great complexity and information content in bioimages, such as the very high density of cells (e.g. astrocytes, microglia, and neurons) intertwined together, or very rapid microtubule growing process in a 4D movie of live cells, it is very challenging to directly apply existing medical image analysis methods to these bioimage informatics problems. Special techniques such as those developed in the FARSIGHT project will be necessary to analyze these complicated image objects. In addition, usually a single biological image stack has a large size (several hundreds of megabytes or even several gigabytes) and several color channels. The objects of interest in such an image, for instance the 3D structures of neurons, could have dramatic variations of morphology and intensity variations from image to image. It is yet not uncommon that thousands of images need to be automatically analyzed in a high-throughput way, in terms of the number of hours or days, but not months or years of manual work. All these difficulties make it necessary to develop novel bioimage informatics algorithms and systems, especially from three aspects: image processing and mining, image database and visualization.

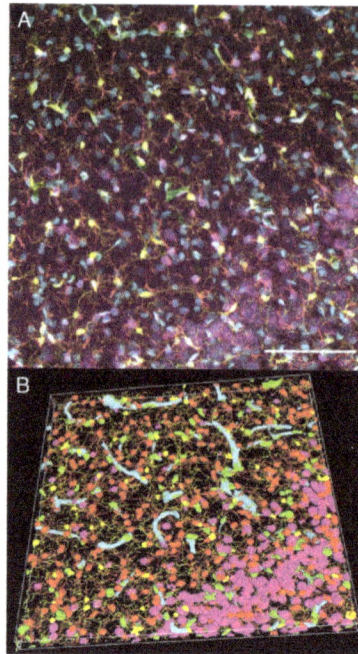

(A) Maximum projection of a 5-channel confocal 3D image of a 100 μm thick section of rat hippocampus. Red: GFAP-labeled astrocytes; green: EBA-labeled blood vessels; yellow: Iba1-labeled microglia; cyan: CyQuant-labeled cell nuclei; purple: NeuroTrace-labeled Nissl substance; scale bar=50 μm. (B) 3D rendering (with a similar color scheme) of the segmented and classified cells produced using the FARSIGHT techniques for (A).

Improving Bioimaging Data during and after Acquisition

While it is well established in the bioimage informatics community that image quality has a dramatic impact on its analysis, the notion of image quality remains quite subjective to many outside the field. In the digital imaging world, the quality of an image is generally measured using mathematical criteria such as the so-called signal-to-noise ratio (SNR), as well as other descriptors that locally characterize the image 'texture' (leading to measures of, for example, contrast or homogeneity). More recently, an image-centric measure known as the SSIM (for structural similarity index measure) was devised to measure the quality of an image in comparison to some reference or 'ground-truth' image (the availability of which is problematic in microscopy). While many of these measures coincide with our visual, human appreciation of what is a 'good' image, they are rather poor indicators of whether the image is actually suitable for subsequent (computerized) analysis. More often than not, the acquisition parameters should be defined and optimized alongside the analysis pipeline itself in order to ensure proper quantification. In some instances, this optimization process may additionally benefit from re-thinking the sample preparation itself, ensuring that the acquisition and analysis are well adapted to the biological question (e.g. a nuclear reporter does not permit optimal cell segmentation, but is better suited for cell counting, notably in dense tissue).

After acquisition, image quality can usually be improved using a range of signal processing methods known as denoising tools. Unfortunately, fluorescence microscopy images often exhibit numerous different artefacts due to the obligatory compromise between image quality and cell viability, and therefore require denoising algorithms that are tailored to each and every case. For instance, excitation of fluorescent dyes can cause toxicity and bleaching, imposing heavy constraints on exposure time and therefore limiting global contrast. Moreover, insufficient excitation (as well as uneven fluorescent labelling) may induce gaps and non-homogeneities within the structures of interest, thus considerably hampering their extraction. Finally, imaging in thick tissue is usually affected by light scattering, causing the SNR to progressively decrease as the illumination source traverses the sample. In most instances, so-called 'pre-processing' algorithms can be used to reduce these artefacts in order to facilitate the subsequent analysis steps.

Commonly used pre-processing methods for tissue imaging can be grouped into one of four categories:

- *Fusion* methods combine data from multiple acquisitions of the same sample, thereby reducing the background noise or the impact of light scattering through the tissue. These methods range from simple line or frame averaging to more complex multi-view fusion of varying sample poses, notably in SPIM.

- *Deconvolution* methods aim at improving image resolution by reducing the artefacts and aberrations introduced by the optical device, defined by its point-spread function (PSF). The PSF not only depends on the imaging technique, but also varies in space and depth, notably in thick or scattering samples, thus raising a significant challenge for the signal processing community.

- *Contrast enhancement* methods are used to homogenize signal distribution over the entire dataset (examples include image normalization and histogram equalization). They can be particularly useful in a three-dimensional context to correct the progressive intensity decrease caused by light scattering in deeper sections of the sample.

- *Filtering* algorithms are used to reduce image noise (locally or globally) using either conventional low-pass or median filters, or by specifically enhancing or correcting heterogeneities of the structures of interest based on local texture information.

Image pre-processing can be of crucial importance to facilitate the subsequent analysis steps, so long as the chosen method is adapted to the specimen and optical device at hand. Nevertheless, it is worth stressing that these techniques apply some transformation that directly affects the pixel values (and thereby the dynamic range) of the original dataset. In some instances, such pre-processed datasets may no longer be usable to extract biological information that directly depends on carefully calibrated intensity values (e.g. protein expression levels). For such sensitive applications, great care should be taken in ensuring that the data are knowingly and coherently adjusted.

Cell and Tissue Segmentation Methods

Image segmentation is one of the cornerstones of digital image analysis, and describes the process of separating an image into different meaningful parts or *segments*. Novel methods are constantly being developed to tackle the new and ever more complex challenges raised by novel or diversifying bioimaging techniques. Cell segmentation therefore remains today one of the most active research topics in the bioimage informatics community, even after decades of investigation.

In the context of tissue analysis, cell segmentation methods fall into one of two categories. In studies concerned with the global localization of cells within tissue (and eventually their lineage throughout morphogenesis), there is only minor interest in extracting the actual shape of each individual cell. Therefore, the cell segmentation problem is reduced to that of finding the cell nuclei, which is arguably simpler in tissue and model organisms where nuclei appear well separated (the problem becomes considerably more complex as tissue density increases). In other studies where either the cell shape or the local neighbourhood is required (for instance to characterize cell–cell interaction and intercalation), the full cell outline must be extracted, and therefore imposes that the cell membranes be imaged. In contrast to nuclear labelling, fluorescent signal is usually less homogeneous along the cell membrane (notably due to poor resolution and light scattering), therefore segmentation is particularly more challenging even after suitable pre-processing, and often requires the help of nuclear localization to initialize the extraction process.

For nuclear segmentation, a straightforward approach lies in intensity-based pixel classification, where each pixel is classified as being part of the nucleus according to some intensity threshold (either global or adapted to the local context). This simplistic approach is fast and efficient on highly contrasted datasets, but quickly becomes error-prone as the SNR decreases. Alternative solutions have been proposed to combine intensity with size thresholds, or use multi-scale analysis perhaps followed by region-growing approaches including watershed and active contours. Nevertheless, a common bottleneck arises in highly dense tissue, where nuclei that do not appear separated are either discarded or merged, leading to under-segmentation. Separating clustered objects is still an open problem with no universal solution, although several specialized methods have been proposed, notably based on watershed, gradient flow, super-pixel grouping or the concept of 'lines-of-sight'.

Cell segmentation in a tissue context is a comparatively much more delicate endeavor, due to the extreme proximity and high density of the microenvironment. Methods in this category fall into one of two categories, depending on their starting point:

- *Region-growing approaches* start by detecting the center of each cell and applying a region-growing approach to reach the cell membrane, which is particularly useful when a nuclear segmentation is already available. However in the absence of a nuclear marker, the central region of the cell can be inferred from local-intensity minima in the membrane signal. Region growing is typically achieved using the watershed approach, where the image is considered as a topographic relief map that is iteratively flooded from every initial seed iteratively until the edges of the water basins meet (defining the so-called watersheds). While the watershed is particularly sensitive to noise, the final membranes are also not always optimally placed, notably on low-resolution datasets where considerable amounts of data are missing. A popular alternative for cell segmentation lies in deformable models (also known as *active contours*), where a contour is initialized around each seed (nucleus) and is attracted towards the cell membrane by forces resulting from the minimization of a cost functional with various parameters controlling the behavior of the contour. Deformable models are more flexible in comparison to other approaches, at the expense of slightly increased computation times, although efficient implementations are available.

- *Direct approaches* rely primarily on the membrane signal, and rather consider the cell membranes as a network that is to be extracted from the image. These approaches typically start from an intensity-based analysis, followed either by a polygonal fitting procedure or morphological operations followed by surface reconstruction and local refinement, not unlike active contours.

The number and diversity of segmentation methods clearly highlight the current lack of a universal, fully automated solution. While each method is usually better suited for a given application, it is worth pointing out the growing interest for machine learning strategies, where the user is allowed to intervene at different steps of the workflow to correct and teach the algorithm when it errs. The growing influence of machine learning in numerous areas of science (and notably in image recognition) is opening a new avenue for future developments in this field, and provides a potential alternative to build a truly generic segmentation tool.

Tracking Cells in Context

Cell tracking is an essential and necessary step towards the understanding of development over time, through different cellular events, such as proliferation, differentiation and migration.

Cell tracking is the process of following the position of each and every cell of interest within its environment over the course of time (typically, a time-lapse acquisition), with the ultimate goal to extract spatio-temporal features such as deformation, migration, intercalation and lineage tree. Similarly to cell segmentation, manual cell tracking is a particularly tedious undertaking, especially on massive three-dimensional sequences comprising thousands of cells and their progeny. Cell tracking has therefore been and continues to be a topic of substantial interest in the bioimage informatics community, as illustrated by the organization of a recent community challenge. Traditionally, tracking methods are classified into two major categories: (i) *association* methods, where all the cells are first segmented in the entire sequence, then the tracks are built from these detections in a subsequent step, and (ii) *evolution* methods, where each cell is sequentially segmented and propagated from frame to frame (typically using template-matching approaches or active contours). In developmental biology, however, the landscape is slightly different. Indeed the computational burden induced by deformable models for segmentation tasks renders their application to tracking even more challenging. Instead, the vast majority of tracking methods rely on association, although in three major ways:

- Frame-to-frame association consists of assigning to each cell in the first frame a cell in the subsequent frame, based on some prior knowledge or hypothesis of motion. The simplest prior is to assign the spatially closest cell in the next frame, but other criteria have been used to improve or refine this association, based either on the global motion of the observed tissue or other information such as shape, intensity and prior knowledge on cell behaviour.

- Graph-based approaches represent the tracking problem in the form of an oriented graph where vertices represent detected cells, while edges represent all the possible links between the vertices over time. The association problem is then solved by a graph-optimization procedure to extract the most plausible temporal pathway of cells through the graph.

- Global optimization strategies have been investigated to segment and track cells simultaneously within the entire spatio-temporal four-dimensional (3D+time) hypervolume in a single step, based on spatio-temporal morphological operators (by analogy with kymograph analysis, this would be equivalent to tracking cells in two dimensions by extracting three-dimensional tubes within a 2D+time hyperstack). Such approaches are efficient on small datasets, but do not scale properly on conventional workstations.

In the vast majority of cases, the efficacy of association-based cell tracking methods is highly dependent on the previous (segmentation) step. A cell that is not properly detected can quickly add to the computational burden of the tracking algorithm, which must rely on more rules and strategies to determine whether the absence of a cell at a given location is due to erroneous segmentation, or rather due to a natural biological process (e.g. cell death, local change in tissue dynamics, etc.). Eventually, excessively strict rules that produce a perfect tracking in a specific situation may inadvertently limit the interpretation of tissue dynamics to only a subset of the (biologically) plausible scenarios. It is therefore crucial to keep a reasonable parameter tradeoff to prevent data over-fitting, which is best solved by returning to and improving the segmentation step.

Towards Reproducible Research Via Open Software Resources

Ingrained in the principles of reproducible research is the verification and reusability of experimental results, facilitating the conception of future experiments and preventing wasting time in reinventing the wheel and/or building upon false conclusions. While the importance of the availability of raw data is now openly recognized, such current practices in the bioimaging field are still far behind other scientific domains, while discussions between journal editors and publishers on developing best-practice guidelines to address this issue are still in their infancy. In recent years however, the bioimage informatics community has been actively pushing to promote reproducibility in both the software and hardware aspects, by releasing analysis algorithms and equipment blueprints online, as well as ground-truth data in the form of publicly available datasets and software for benchmarking purposes.

Several such developments have directly targeted the developmental biology community. Each solution has its own strengths and weaknesses, which generally stem from the biological application driving its development. For instance, general-purpose software platforms typically provide a wide range of analysis tools to analyse bioimaging data (including tissue), although they may exhibit lower efficacy in specific contexts where specialized (though less generic) solutions have been developed. While we do not provide an exhaustive list of such solutions, we focus here on some of the most recent or active developments in the field of epithelial and embryonic development, most of which are available as open-source platforms:

- Icy and ImageJ/Fiji are typical illustrations of such general-purpose, community-driven bioimaging platforms providing a large variety of tools and frameworks for data acquisition, visualization and analysis in biology, with a particular focus on extensibility via community-contributed plugins, scripts and protocols. They have been successfully applied to developmental biology studies through various extensions. For example, one such extension for the Icy platform studies the spatial organization of the embryonic mouse heart, by combining a filtering approach to increase membrane signal acquired in confocal microscopy, and a segmentation tool based on deformable models to extract both the nuclei and the cell membranes.

- MARS-ALT is among the first reported freely available software dedicated for the segmentation and lineage tracing of cells in three-dimensional time-lapse imaging data for developmental biology studies, and more specifically in the context of plant growth in the *Arabidopsis thaliana* model plant. The pipeline starts with an algorithm to fuse multi-angle confocal three-dimensional imaging stacks. The cells are then segmented using the watershed algorithm, and tracked over time using nonlinear registration of the consecutive time points, followed by a graph-based matching algorithm. It is worth pointing out that this software, although freely available, is not open-source, and therefore does not fully comply with reproducible research principles.

- Automated cell morphology extractor (ACME) is software developed for the automated reconstruction of cell membranes in three dimensions. The software provides an all-in-one solution for dense tissue that sequentially applies data-filtering for membrane structure enhancement, followed by the segmentation of cells in dense tissues. It was specifically developed for zebrafish data (illustrated on neuroectoderm and paraxial mesoderm data) obtained via confocal and two-photon microscopy.

- Focusing more particularly on the dynamics of epithelial tissue during development, the EDGE4D software has been proposed to segment and track cells across the complex process of epithelial folding during *Drosophila* gastrulation. The algorithm combines nuclear and membrane signals to provide superior segmentation to its aforementioned counterparts in the challenging context of multi-photon microscopy.

- MorphoGraphX is a platform developed for the quantification of four-dimensional (three-dimensional time-lapse) analysis of growth dynamics in plant morphogenesis, focused on computational efficiency and realistic visualization through the use of graphics processing units (GPU). This software provides specific routines to segment and track the outer cell layer of the embryo (represented as an unstructured graph), by leveraging the power of existing open-source packages including CImg and ITK. It has mainly been applied in the context of *Arabidopsis*, but has been shown to work with other models, notably *Drosophila*.

- TissueMiner is a framework to quantify epithelial cell dynamics in living tissues over prolonged acquisitions. It specifically includes tools to generate a database from cell segmentation and tracking results, and provides a number of quantitative spatio-temporal descriptors of epithelial morphogenesis, and was applied to study the *Drosophila* pupil wing. The interface is based on the R platform and provides utilities to query and visualize the database in an efficient manner.

- Real-time accurate cell shape extractor (RACE) is a high-performance image analysis framework for automated three-dimensional cell segmentation, with a particular focus on massive SPIM datasets. It provides the necessary software to extract nuclei and cell shapes from terabyte-sized images, and therefore constitutes a step towards scalability and high-throughput studies, as illustrated by examples on *Drosophila*, zebrafish and mouse embryos. Additionally, a companion tracking software was developed to study large-scale embryonic development.

- EpiTools is another open-source toolkit for the study of developing epithelial tissue. Based on Matlab, it provides graphical interfaces to segment and track the contour of cells in 3D+time confocal series of membrane-labelled cells. It also closely integrates with the Icy platform by providing specific extensions to open and visualize the segmentation and tracking results within Icy. The tool is specifically developed to study the dynamics of epithelial growth and morphogenesis, and has been applied to the *Drosophila* wing disc.

Biodiversity Informatics

Biodiversity informatics is a rather new field, which can be defined as the creation, improvement, analysis, and interpretation of information regarding biodiversity. The field was born out of the longtime information-management of the systematics community, in tandem with the realization of the desperate situation of the 'biodiversity crisis' in recent decades. Although biodiversity information has been accumulated for centuries, this information has not been managed effectively until very recently. Over the past two decades however, large-scale efforts have concentrated on the challenge of building, improving, and understanding these information resources; what we now term the field of biodiversity informatics.

Biodiversity informatics plays a central enabling role in the research community's efforts to address scientific conservation and sustainability issues. Great strides have been made in the past decade establishing a framework for sharing data, where taxonomy and systematics has been perceived as the most prominent discipline involved. To some extent this is inevitable, given the use of species names as the pivot around which information is organised. To address the urgent questions around conservation, land-use, environmental change, sustainability, food security and ecosystem services that are facing Governments worldwide, we need to understand how the ecosystem works. So, we need a systems approach to understanding biodiversity that moves significantly beyond taxonomy and species observations. Such an approach needs to look at the whole system to address species interactions, both with their environment and with other species.

It is clear that some barriers to progress are sociological, basically persuading people to use the technological solutions that are already available. This is best addressed by developing more effective systems that deliver immediate benefit to the user, hiding the majority of the technology behind simple user interfaces. An infrastructure should be a space in which activities take place and, as such, should be effectively invisible.

Grand Challenge

The grand challenge for biodiversity informatics is to develop an infrastructure to allow the available data to be brought into a coordinated coupled modelling environment able to address questions relating to our use of the natural environment that captures the '*variety, distinctiveness and complexity of all life on Earth*'.

Biodiversity processes are complex and can have a large, long-term impact at the macro-scale, even if they have occurred rapidly at the sub-cellular molecular level e.g., the Phosphate cycle. Processes taking place in seconds over scales of nanometres, are crucial to processes that take years and at scales of many hectares and ultimately to planetary processes in geological time. Capturing such inter-dependent processes, across such a breadth of scales, is beyond the capability of current information management and modelling methods. To have an impact on biodiversity conservation, sustainability or our environment, we need to consider all aspects of biodiversity, from genes to ecosystems, in a holistic approach. We need to be able to assess global biodiversity changes and make predictions about ecosystems. We need to be able to integrate different facets of past and present environmental and biodiversity observations and embed them in models with predictive power. We will need to develop new models to address socially urgent questions. Such an approach will take biodiversity science far beyond a collection of taxon names, capturing data about different facets of biodiversity, both by their absolute position and their relative position, together with their observational and temporal context. Most importantly, through biodiversity informatics, biodiversity scientists will be able to understand, to measure and predict how change affects the actual functioning of the ecosystem.

Need for Biodiversity Informatics

Much of the biodiversity information is available in the form of scientific collections such as specimens, herbaria, microorganism repositories in various universities, natural history museums, research institutions and organizations concentrated mainly in developed countries. Experts are often asked for quick advice or input by government and private agencies regarding issues such as status of a species population in a particular region or area, potential effects of introduced species, forest fire impacts in a protected area, ecological effects of development, and so on. Such experts are often hamstrung by the lack of easily accessible information.

Therefore, the need of the hour is to digitize all the available information on biodiversity and place it on information networks to aid the current generation of researchers.

Many international conservation information networks have been developed during the last few decades, most of which are open to public. These information networks include news, expertise, searchable databases, and any other kind of relevant material that can be put on websites.

Some major databases providing online information services for conservation		
Name	Website	Description
GBIF	https://www.gbif.org/ Hotspots	Global biodiversity facility, unit level records
Hotspots	https://www.conservation.org	Provides information on global hotspots
ETI-WBD	http://www.eti.uva.nl/tools/wbd.php	Global taxonomic database
IUCN SIS	https://www.iucn.org/themes/ssc/our_work/sis.htm	IUCN Information service on species
IUCN Redlist	http://www.iucnredlist.org/	IUCN information on conservation status of species
UNEP-WCMC	https://www.unep-wcmc.org/	Information centre with multiple databases on conservation

Biodiversity networks and databases will play a crucial role in managing the vast and increasing information on biodiversity components all around the world.

The databases are playing an important role in providing baseline as well as key information to many scientists and researchers working for conservation projects in different regions of the world. They have also inspired others to develop databases and information networks for ease of information dissemination and networking in their area of expertise.

The information networks may be regional or global and may be complex or fairly simple in structure depending upon their objectives. Nevertheless, experts involved in constructing these information networks tend to make them user-friendly for public convenience.

At the global level, GBIF is playing a lead role and is the most ambitious of all information networks as far as integration of biodiversity data is concerned. GBIF's data portal is already available online as a standard source providing integrated primary data (label or observational) on all species of living organisms. It does this by providing software and web services (using extensible markup language, or XML) that enable integration of data drawn from multiple resources distributed around the world.

GBIF is currently serving over 10 million records from 34 distributed databases and intends to make the entire world's biodiversity information available to all within the next 10 year period. The GBIF mission statement is "to make the world's primary data on biodiversity freely and universally available via the Internet". Currently, 47 countries and 32 international organizations are members in GBIF, and the membership steadily grows, as does the data content. Ocean Biogeographic Information System (OBIS) is among the largest data providers to the GBIF.

Applications of Biodiversity Informatics to Biogeographical Questions

Some applications of this new conjunction of data and methodologies focus on issues of basic science, like the study of evolutionary processes, causes of range limitation or species' reactions to changing environments. In Peterson, niches were modelled for 37 pairs of species. This was done by using a genetic algorithm to search for regions in the map that are 'similar', in terms of annual precipitation, average temperature, elevation and potential vegetation to those regions where the species has been reported. Species presence data comes from extensive museum databases. The hypothesis was that related taxa should share niche features, thus confirming theoretical predictions about niche conservatism. Indeed, that was found, to a high degree of statistical significance, by reciprocal predictions among related and unrelated pairs of species. In a climate change-related application the fundamental ecological niches of 1870 species of Mexican birds, mammals and butterflies were estimated using G_{arp} again, and the resulting niches were projected to future climates, obtained from general circulation models. Several analyses were then performed on likely changes of distribution areas under several scenarios of dispersal capabilities. The results highlight the relevance of mountain chains for conservation, as turnover of species is lower in mountainous areas than in the central plains of Mexico.

Still other applications focus on management issues, like biodiversity exploration or location of protected areas, assessment of the potential for pest damage to crops or evaluation of possible routes for invasive species or diseases, to name just a few examples.

To provide an example of analysis of an invasive species, considerable recent concern has been caused by the cactus moth Cactoblastis cactorum, which as an invader could be catastrophic to certain cactus species, particularly the Platyopuntia. We drew location data for C. cactorum from scientific collections of the Smithsonian Institution, and used them to estimate hemispheric niche dimensions in terms of climatic variables. The geographical display of regions of high similarity (on the basis of the climatic variables chosen) to those where the species has been observed provides a prediction of the potential distribution for the species in North America. Then, geographical distributions of species of Platyopuntia cacti were obtained by first modelling their niches using the Garp algorithm via 5099 observational data provided by several herbaria. The niches so obtained were then reduced by biogeographical reasoning supervised by experts on the group. Those two procedures yielded individual distributional ranges for 60 species of Platyopuntia on the North American continent. Overlay of Cactoblastis' niche with the distributions of its host plants thus provided a first approximation to understanding the potential route of invasion by Cactoblastis into the deserts of the southwestern USA and northern Mexico.

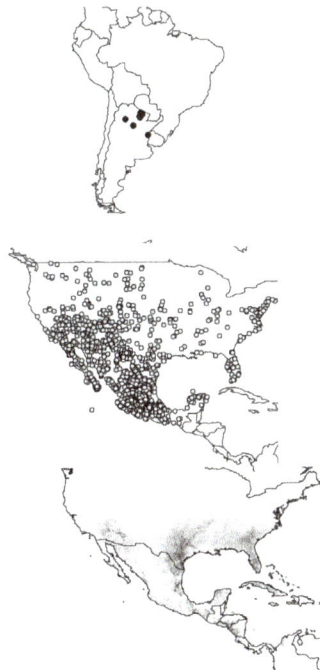

Description of the figure above- (a) Localities of *Cactoblastis cactorum* specimens amenable to precise georeference. Specimens in the Smithsonian Natural History Museum. (b) Localities for 5099 specimens in the subgenus *Platyopuntia* in Mexico and the USA. (c) Bio- climatic surface, based on the *C. cactorum* specimens and calculated with the FloraMap software based on a principal component analysis involving three environmental variables distributed over a 12-month period.

Challenges and Limitations

Although of great potential, significant challenges do exist for this new world of BI. In the first place, presence data include significant biases in the spatial and temporal distribution of collecting efforts and in its overall quality. The dynamic nature of taxonomy means that databases that are not maintained actively may soon be outdated, with synonyms comprising 10–30% or more of names in many databases. The ageing nature of many collections, owing to inattention or to lack of recent material, makes collections data challenging to interpret in light of ongoing land use changes. Hence, in a time when gigabytes of primary biodiversity information are becoming available to all, issues related to quality control are more crucial than ever.

Although the problems mentioned above can be found in many biological databases, the heterogeneous origin of Web-assembled databases makes quality control even more important. As the origin of the data is heterogeneous, record quality may be uneven and numerous procedures must be used to detect and correct problems. Some of the more common problems include the following.

Specimens may have wrong identifications: This error is quite frequent, and yet can be extremely difficult to detect and correct. Without expert participation in inspection and determination of the original specimens, only very obvious mistakes will be detected. Obviously, data from poorly determined collections should be used only with care when developing biodiversity analyses. More generally, records from such collections should be flagged clearly or perhaps even not opened to search and query by nonprofessionals.

Outdated taxonomy: An additional suite of problems arises from the evolving nature of biological taxonomy: species identified correctly at one point in time as species X may later be assigned a different 'correct' name; splitting one species into several, changing generic affinities, etc., all may create such situations. Consultation of taxonomic authority files combined with geographical information about the geographical distributions of species to which those names refer, may permit identification of such names. For example, in preparation of the C. cactorum example, we made use of a small database (5099 records) of Opuntia cacti drawn from several institutions in the USA and in Mexico. A considerable number of specimens were listed under outdated names (e.g. O. schotti, which is now considered as Grusonia schotii (Engelm.)). A first check against an updated authority file for the Cactaceae of Mexico detected many such inconsistencies; detailed geographical inspection was necessary to detect other problems, leaving the database relatively clean taxonomically. Unfortunately, in general, obtaining reliable, updated taxonomic authority files is a major problem for most taxa. The large taxonomic information services (e.g. Species 2000, The Integrated Taxonomic Information Service, Missouri Botanical Garden's Tropicos, the Index Kewensis and others) remain far from complete—their completion will remove one major obstacle for proper quality control of distributed specimen databases.

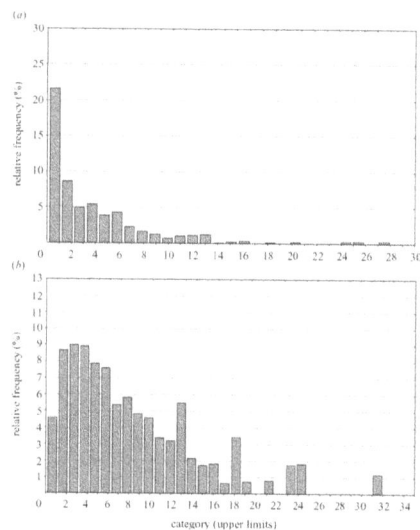

Distribution number of synonyms in two catalogues of Mexican species. (a) Poaceae (b) Cactaceae.

Faulty georeferencing: Frequently, the identification and textual description of the collection locality may be correct but the geographical coordinates assigned to that site may be erroneous. Faulty georeferencing can be detected by means of consistency analyses, in which verbal descriptions of locality are checked against the geographical coordinates. At present, only a small minority of localities in museum databases are properly georeferenced, which, of course, raises the more basic question of how to add georeferences to specimen data quickly and efficiently. One important example is that of the MaNIS project, a community effort to integrate and georeference data from mammalian specimens in 17 museums: out of the 296 737 localities in the original pool of localities, only ca. 92 000 localities still remain to be georeferenced; the rate of advance is ca. 12 specimens per hour (J. Wieczorek, personal communication) The National Commission on Biodiversity of Mexico has obtained about two million georeferenced specimens, either georeferencing in-house or by cooperation with taxonomists and experts in museums and herbaria.

About 70–80% of specimen label data can be georeferenced by simple techniques and the use of gazetteers; the remaining localities may either prove impossible to reference or feasible only via the participation of experts familiar with the actual collectors. Recently, and most interestingly, much of this process is becoming automated in projects like BioGeoMancer- recognition of locality strings has been made 'smarter', and interpreted locality descriptors are then compared with national or worldwide gazetteer databases; in this way, the bottleneck steps in the georeferencing process are automated, and human participation is focused more at the level of supervision and error checking, making the process considerably more efficient.

In conclusion, the massive storehouse of distributed, raw biodiversity data that the Internet is enabling will set the stage for how biodiversity patterns are analyzed in the future. Abundant examples already demonstrate the rich potential of such data when analyzed and interpreted in the context of geospatial information as part of the nascent field of BI. Nevertheless, the demands that these technological advances will put on the shoulders of the taxonomic and systematics communities are significant—in fact, without a strong and active taxonomic community, BI will never be more than a clever set of software tools lacking a substantial factual basis.

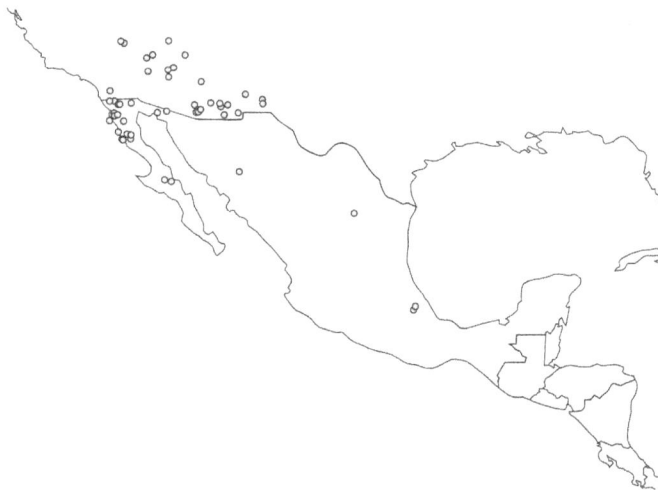

A map of registers for Opuntia chlorotica was presented to a specialist. The specimens for the obvious outliers in southeastern Mexico were checked and found to be misidentifications by the specialist.

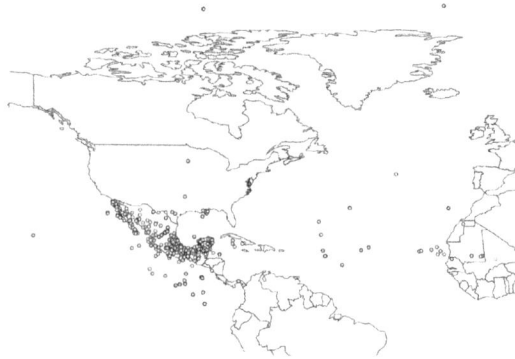

All the points depicted in this figure are specimens of terrestrial plants of Mexico with a faulty geo-referencing. Some of them are obvious, like the Polar, mid-Atlantic or African ones. Others require careful comparison with standardized maps of Mexico, with accepted polygons for states and municipalities (often there are conflicts between states on the precise location of their borders), and a scale, projection and datum that allow consistent comparisons.

Detection of problems associated with synonyms, misidentifications, georeferencing inconsistencies, outdated taxonomy, and so on depend on the existence and enthusiastic participation of an active community of taxonomists. More importantly, these advances depend critically on adequate support to the fundamental infrastructure of the museums and herbaria of the world— these institutions provide the key infrastructure of the world of knowledge regarding biodiversity and every day are more endangered by cost-cutting bureaucrats.

Of course, we hope that the exploration of world biodiversity will continue and will gain new strength. Recent promising initiatives, unfortunately, have not achieved full success. For example, the most recent attempt to discover and describe the remaining species, the All Species initiative, was delayed owing to the change in world economy. The Global Taxonomy Initiative of the Convention on Biological Diversity was launched, but has not been funded. Failure to carry through with these initiatives is worrisome, as much remains to be explored, and many critical elements of biodiversity remain to be discovered.

It is clear that 'DNA Taxonomy' will add speed to the exploration of biodiversity, although there is still debate about how many of the claims of its proponents are real or feasible. In any case, one major lesson learned from specimen-based BI analysis is that proper vouchering and georeferencing of specimens is the sine qua non of macroecological and biogeographical analysis. Simply 'DNA barcoding' specimens, without precise (as precise as possible, and with modern technology this is metres) reference to the locality may serve to tackle many problems in systematics, evolutionary biology and other fields, but will probably leave out whole categories of analysis.

Biodiversity Informatics is adding value to taxonomic activity in ways probably not foreseen even 10–20 years ago, and is becoming indispensable in developing countries striving to manage their biodiversity adequately. Indeed, in developing countries, national taxonomic institutions are often small and under-funded. Many large countries—and particularly those that qualify as 'megadiverse' —are nevertheless essentially unexplored for many taxonomic groups. For such countries, a practical answer to the lack of national taxonomic efforts or institutions is to refer to existing information and knowledge, using the array of techniques described above to improve insight. If or when the wealth of information that is held in world natural history museums and herbaria is available efficiently to those countries, the way biodiversity is managed will change radically.

To this end, enormous activity is required on the part of the museums and herbaria. Requisites include the following.

(i) Museums and herbaria must continue the enterprise of collecting new biodiversity material, which should be the richest in information content (e.g. precise geographical coordinates, detailed digital imagery or sound recordings, DNA profiling). In general, our biodiversity resources are ageing, and not only are new phenomena not represented, but a baseline of highest-quality information associated with recent specimens is in general lacking.

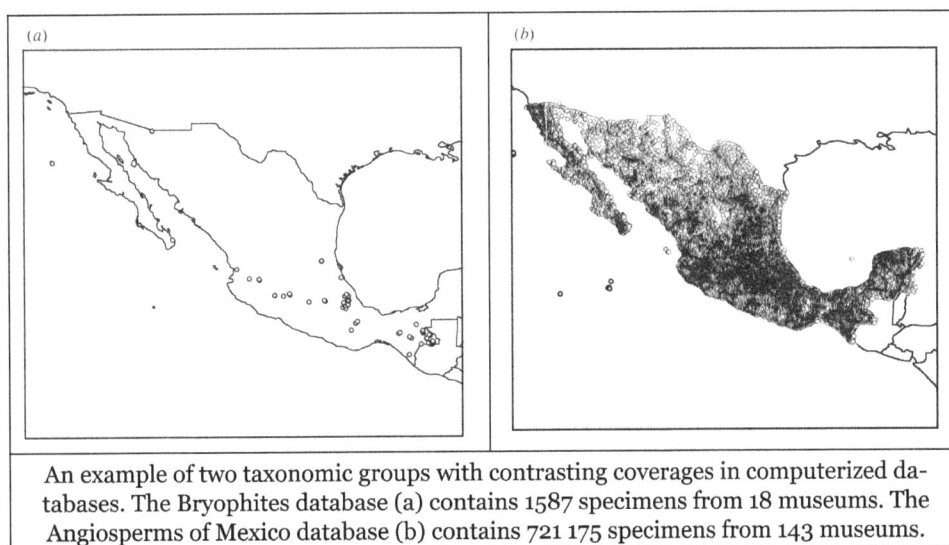

An example of two taxonomic groups with contrasting coverages in computerized databases. The Bryophites database (a) contains 1587 specimens from 18 museums. The Angiosperms of Mexico database (b) contains 721 175 specimens from 143 museums.

(ii) The museums must increase the pace for releasing good-quality raw data. Good quality means well determined (low misidentification rates), georeferenced and quality assessed and corrected (inconsistency checks performed). The cost of the above steps is significant, on the order of US $1– 10 per specimen, without counting the base costs of building, curating and preserving the collections, as well as just running the institutions. The creation of the GBIF, with the purpose of promoting that the world's biodiversity data will become freely and universally available, should take the preceding efforts a step forward by providing technical and financial support and by leveraging national resources and commitment to its objective.

(iii) Investment in development of user-friendly, industrial-strength analytical tools should be increased. Current tools are in the 'artisan' stage, requiring suites of programs and file types to perform a single analysis, and often requiring hours and even days of computer time. Many existing software programs cannot handle the huge data matrices involved in the new world of BI, with its high spatial resolution and multispecies analyses. An effort of tool development similar to that applied to genomic and proteomic informatics will be required. Eventually, efforts will have to start to hasten the convergence of the two 'bioinformatics': the one oriented to suborganismic levels, and the one associated to species and ecosystems. Without doubt DNA barcoding will provide a powerful link between those two fields.

(iv) Training in BI will be a must. University curricula in this field are currently lacking—what is needed is a melding of aspects of biology, computer science and geography, a combination that is not often considered in university programmes.

In summary, BI consists of much more than the particular databases, tools and applications that we have mentioned in this review. Rather, BI is a sweeping, fundamental area of inquiry of which present analyses have touched only the smallest part. Many exciting and far-reaching innovations and steps forward remain to be developed, which will open new doors to funding, activity and further discovery.

Along with the novelty of the field are challenges as well. Assuring that appropriate credit is given where due (e.g. to collectors and curators), and protecting institutional ownership rights to data and their descendent products, may represent significant complications for the development of this field. The BI community will have to develop and adopt the rules of behavior that enhance the sharing of data, while preventing the proliferation of free-riders.

More fundamentally, these developments will involve the evolution of taxonomy and systematics beyond their traditional borders. Instead of just producing the traditional systematic revisions and phylogenetic treatments, BI activities will increasingly draw taxonomists and systematists into analyses and studies focused on issues of interest to agriculture, public health, invasive species and conservation. Although a distraction from the usual tasks of taxonomists and systematists, these issues are nevertheless key in developing BI into a field that will make the case for continuing and increasing support for the taxonomic and systematic enterprise.

Neighbor Joining

The neighbor-joining method is a distance based method for constructing evolutionary trees. It was introduced by Saitou and Nei, and the running time was later improved by Studier and Keppler. It has become a mainstay of phylogeny reconstruction, and is probably the most widely used distance based algorithm in practice. With a running time of $O(n^3)$ on n taxa, it is fast for small input, and empirical work shows it to be reasonable accurate, at least for cases where the rate of evolution is not extremely high or low. St. John et al. even suggest it as a standard against which new phylogenetic methods should be evaluated.

The neighbor-joining method is a greedy algorithm which attempts to minimize the sum of all branch-lengths on the constructed phylogenetic tree. Conceptually, it starts out with a star-formed tree where each leaf corresponds to a species, and iteratively picks two nodes adjacent to the root and joins them by inserting a new node between the root and the two selected nodes. When joining nodes, the method selects the pair of nodes i, j that minimizes the branch-length sum of the resulting new tree. One way of achieving this is always to select the pair of nodes i, j that minimizes

$$Q_{ij} = (r - 2)d_{ij} - (R_i + R_j),$$

where d_{ij} is the *distance* between nodes i and j (assumed symmetric, i.e., $d_{ij} = d_{ji}$), R_k is the *row sum* over row k of the distance matrix: $R_k = \sum_i d_{ik}$ (where i ranges over all nodes adjacent to the root node), and r is the *remaining* number of nodes adjacent to the root. When nodes i and j are joined, they are replaced with a new node, A, with distance to a remaining node k given by

$$d_{Ak} = (d_{ik} + d_{jk} - d_{ij})/2.$$

This formulation of the neighbor-joining method gives rise to a canonical algorithm that performs a search for $min_{i,j} Q_{ij}$, using time $O(r^2)$, and joins i and j, using time $O(r)$ to update d. This search and join is continued until only three nodes are adjacent to the root (i.e. for n - 3 joins where n is the total number of species). The total time complexity becomes $O(n^3)$, and the space complexity becomes $O(n^2)$ (for representing the distance matrix d).

Negative Branch Lengths

As the neighbor-joining algorithm seeks to represent the data in the form of an additive tree, it can assign a negative length to the branch. Here the interpretation of branch lengths as an estimated number of substititions gets into difficulties. When this occurs it is adviced to set the branch length to zero and transfer the difference to the adjacent branch length so that the total distance between an adjacent pair of terminal nodes remains unaffected. This does not alter the overall topology of the tree.

Neighbor Joining Method

- Advantages
  - is fast and thus suited for large datasets and for bootstrap analysis
  - permist lineages with largely different branch lengths
  - permits correction for multiple substitutions
- Disadvantages
  - sequence information is reduced
  - gives only one possible tree
  - strongly dependent on the model of evolution used.

Example of the Method

Suppose we have the following tree:

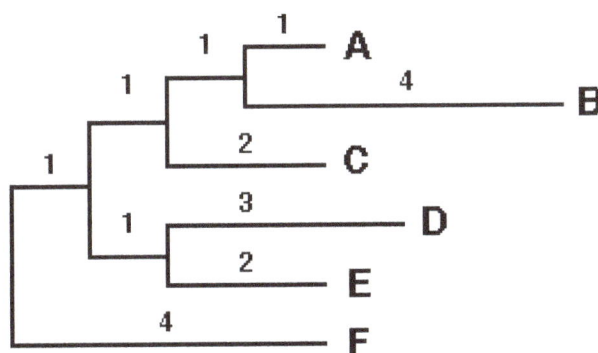

Since B and D have accumulated mutations at a higher rate than A. The Three-point criterion is violated and the UPGMA method cannot be used since this would group together A and C rather than A and B. In such a case the neighbor-joining method is one of the recommended methods.

The raw data of the tree are represented by the following distance matrix:

	A	B	C	D	E
B	5				
C	4	7			
D	7	10	7		
E	6	9	6	5	
F	8	11	8	9	8

We have in total 6 OTUs (N=6).

Step 1: We calculate the net divergence r (i) for each OTU from all other OTUs

r(A) = 5+4+7+6+8=30

r(B) = 42

r(C) = 32

r(D) = 38

r(E) = 34

r(F) = 44

Step 2: Now we calculate a new distance matrix using for each pair of OUTs the formula:

$M(ij) = d(ij) - [r(i) + r(j)] / (N-2)$ or in the case of the pair A,B:

$$M(AB) = d(AB) - [(r(A) + r(B)] / (N-2) = -13$$

	A	B	C	D	E
B	-13				
C	-11.5	-11.5			
D	-10	-10	-10.5		
E	-10	-10	-10.5	-13	
F	-10.5	-10.5	-11	-11.5	-11.5

Now we start with a star tree:

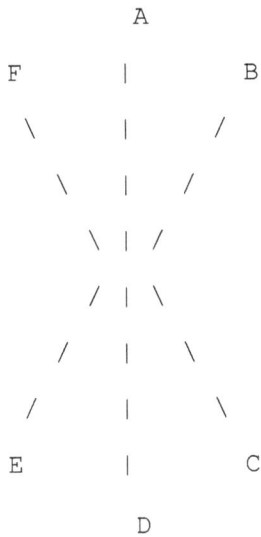

```
          A
F         |         B
 \        |        /
  \       |       /
   \      |      /
    \  |  /
    /  |  \
   /   |   \
  /    |    \
 /     |     \
E      |      C
       |
       D
```

Step 3: Now we choose as neighbors those two OTUs for which Mij is the smallest. These are A and B and D and E. Let's take A and B as neighbors and we form a new node called U. Now we calculate the branch length from the internal node U to the external OTUs A and B.

```
S(AU) =d(AB) / 2 + [r(A)-r(B)] / 2(N-2) = 1
```

```
S(BU) =d(AB) -S(AU) = 4
```

Step 4: Now we define new distances from U to each other terminal node:

```
d(CU) = d(AC) + d(BC) - d(AB) / 2 = 3
```

```
d(DU) = d(AD) + d(BD) - d(AB) / 2 = 6
```

```
d(EU) = d(AE) + d(BE) - d(AB) / 2 = 5
```

```
d(FU) = d(AF) + d(BF) - d(AB) / 2 = 7
```

and we create a new matrix:

	U	C	D	E
C	3			
D	6	7		
E	5	6	5	
F	7	8	9	8

The resulting tree will be the following:

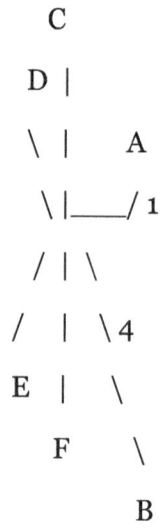

```
        C
      D |
       \ |    A
        \|___/ 1
       / | \
      /  | \4
     E  |   \
        F    \
              B
```

N= N-1 = 5

The entire prodcedure is repeated starting at step 1.

Permissions

We would like to thank the editorial team for lending their expertise to make the book truly unique. They have played a crucial role in the development of this book. Without their invaluable contributions this book wouldn't have been possible. They have made vital efforts to compile up to date information on the varied aspects of this subject to make this book a valuable addition to the collection of many professionals and students.

This book was conceptualized with the vision of imparting up-to-date and integrated information in this field. To ensure the same, a matchless editorial board was set up. Every individual on the board went through rigorous rounds of assessment to prove their worth. After which they invested a large part of their time researching and compiling the most relevant data for our readers.

The editorial board has been involved in producing this book since its inception. They have spent rigorous hours researching and exploring the diverse topics which have resulted in the successful publishing of this book. They have passed on their knowledge of decades through this book. To expedite this challenging task, the publisher supported the team at every step. A small team of assistant editors was also appointed to further simplify the editing procedure and attain best results for the readers.

Apart from the editorial board, the designing team has also invested a significant amount of their time in understanding the subject and creating the most relevant covers. They scrutinized every image to scout for the most suitable representation of the subject and create an appropriate cover for the book.

The publishing team has been an ardent support to the editorial, designing and production team. Their endless efforts to recruit the best for this project, has resulted in the accomplishment of this book. They are a veteran in the field of academics and their pool of knowledge is as vast as their experience in printing. Their expertise and guidance has proved useful at every step. Their uncompromising quality standards have made this book an exceptional effort. Their encouragement from time to time has been an inspiration for everyone.

The publisher and the editorial board hope that this book will prove to be a valuable piece of knowledge for students, practitioners and scholars across the globe.

Index

www.ingramcontent.com/pod-product-compliance
Lightning Source LLC
Chambersburg PA
CBHW082034190326
41458CB00010B/3359